PREPARE, RESPOND, RENEW

APPLYING GIS

PREPARE,

RESPOND,

RENEW

GIS FOR WILDLAND FIRE

Edited by
Anthony Schultz
Matt Ball
Matt Artz

Esri Press
REDLANDS | CALIFORNIA

CONTENTS

INTRODUCTION

WILDFIRES CLAIM LIVES, DESTROY STRUCTURES, AND devastate communities and landscapes. The increasing areas where development meets nature—and more days of hot and dry weather—have magnified the impact of wildfires from Canada to Australia and around the world. The response to and recovery from increasingly complex firestorms stress budgets, economies, communities, and environments. In California, for example, the cost of wildfire destruction has prompted some insurance companies to stop writing new homeowner policies.

Saving lives, preserving property, and protecting the environment are the primary responsibilities of governments and many organizations. Achieving these goals requires collaboration across agencies, organizations, industries, and homeowners. Wildfire management requires an integrated approach to build community and environmental resilience.

Fast-moving fires, such as the deadly windblown blaze in Maui, Hawaii, in 2023, create an array of challenges for firefighters, police, homeowners, insurance companies, relief agencies, utilities, and others. Increasingly, responders use the latest tools of geographic information systems (GIS) to analyze wildfires through data that can be modeled to visualize threats in real time. Preemptively, GIS helps firefighters model how wildfires spread depending on weather, geologic features, and human development. Through predictive analytics and mapping technologies, firefighters can model the direction and rate of spread of wildfire to give a community, a nature preserve, a fire department, or a single homeowner time to prepare for, or even prevent, the next wildland fire. As they identify priorities and reduce fire vulnerabilities, agencies can visualize, record, and track the status of their accomplishments in the field.

GIS capabilities

GIS can address any part of the wildfire life cycle—from the time a fire erupts, grows, and develops fully until it burns out or is extinguished—followed by new growth. GIS supports fire preparedness, mitigation, response, and recovery and rehabilitation. GIS tools help firefighters identify, analyze, and understand landscapes for wildland fire management and protection.

GIS has supported the wildland fire community for decades and continues to be used in an effective collaboration across governments, organizations, and industries. More recently, GIS has incorporated such tools as AI, VR, predictive analysis, and drone imagery to support the application of location-based analytics. Agencies can gain greater insights using contextual tools to visualize and analyze data; collaborate with others; and share information through maps, apps, imagery, and reports.

Spatial analytics

Spatial analytics is at the center of ArcGIS® technology in identifying areas prone to wildfire, planning for smarter communities, and preparing for potential postfire threats, such as landslides and flooding in burn areas.

Imagery and remote sensing

GIS helps fire agencies and other organizations manage and extract answers from imagery and remotely sensed data. It includes tools and workflows for visualization and analysis and access to the world's largest imagery collection.

Mapping and visualization

Maps identify spatial patterns in the data, resulting in better decisions and strategies. Maps also facilitate collaboration among sometimes disparate agencies, organizations, and community groups by providing a common operating picture of the fire threat.

Real-time GIS

The ArcGIS suite supports real-time monitoring of wildfires from any type of sensor or device—useful in creating, using, and sharing maps; accelerating response times; optimizing safety; and identifying available assets for fire response.

GIS in 3D

GIS brings real-world context to wildfire planning and response, turning data into smart 3D models and visualizations to reveal patterns, share ideas, and solve problems.

Data collection and management

GIS allows users to collect, crowdsource, store, access, and share data efficiently and securely, as well as integrate data in their information systems and geoenable any type of data from any source.

Stories and strategies

Prepare, Respond, Renew: GIS for Wildland Fire presents a collection of real-life stories that illustrate how organizations use GIS to solve challenges and create better outcomes. The stories share how organizations have integrated GIS, a geographic approach, and spatial reasoning into all aspects of wildland fire management. The book concludes with a section about getting started with GIS and suggests strategies to build location intelligence into decision-making and operational workflows. The book presents location intelligence as a layer of knowledge that practitioners can add to their existing experience, expertise, and perspective and incorporate into their daily operations and planning.

If your organization is not currently using a geographic approach in its decision-making or daily operations, this book can be used to start developing skills in those areas. Developing these skills does not require deep GIS expertise, nor does it require professionals to disregard all their experience and knowledge. Using a geographic approach is another way to think about solving problems in a real-world context.

HOW TO USE THIS BOOK

THIS BOOK IS DESIGNED TO HELP YOU FOCUS ON ISSUES that matter to you right now. It is a guide for taking first steps with GIS and applying location intelligence to decisions and operational processes to solve common problems and create a more collaborative environment in your organization. You can use this book to identify where maps, spatial analysis, and GIS apps might be helpful in your work and then, as next steps, learn more about those resources.

Learn more about GIS resources for wildland fire management by visiting the web page for this book:

go.esri.com/prr-resources

EL DORADO

PART 1

PREPAREDNESS

A GEOGRAPHIC APPROACH TO WILDFIRE PREPAREDNESS planning helps communities and authorities plan for the inevitability of wildfires and attempt to reduce the risks. Through predictive mapping and analysis, agencies can identify, prioritize, and implement risk reduction projects to reduce a community's vulnerabilities. This data-driven process underpins community planning and the process for securing needed resources. Wildfire preparedness maps allow agencies and local authorities to visualize current risks and monitor historical trends that can inform a real-time approach to changing conditions and potential impacts on residents, infrastructure, and the environment.

Many agencies use ArcGIS software tools to plan for and respond to wildland fires. For example, many community wildfire protection plans (CWPPs) use a template that incorporates ArcGIS Hub℠ technology to organize agencies, communities, and organizations around wildfire planning efforts. Hub includes a variety of web maps and apps that help communities evaluate the risk to critical infrastructure, residences, and the environment. Hub also acts as a digital town square in which residents and other stakeholders can monitor progress on projects that reduce wildfire risk and provide dates and locations of upcoming meetings.

Understand your risk

Fire staff can use GIS to view and understand physical features and the relationships that influence fire behavior. They can view factors such as topography, fuel moisture, and vegetation type to determine locations with the highest fire risk. They can compare this information with high-value resource locations such as wildlife habitat and ecosystems, infrastructure, cultural resources, sensitive soils near drainages, and housing development to pinpoint areas at greatest risk. Command staff can determine the likelihood of wildfire occurrence by locating historical fire locations and identifying potential ignition sources such as power lines, roads, industrial areas, and housing. When areas of high potential threat are overlaid near flammable vegetation and valuable resources, areas in need of risk reduction can be identified for risk mitigation projects.

Identify trends

Fire staff can use the results of a wildfire risk analysis to develop a comprehensive plan. GIS can help map and analyze where priority fire prevention, vegetation management, and wildfire detection programs are needed. The analysis can also support initial fire response plans for at-risk areas before a fire starts. Maps aid evacuation modeling by defining, understanding, and anticipating population, demographics, social inequities, and traffic patterns that can support a safe and efficient evacuation.

Develop a plan

Using GIS, agency staff can work in the field to collect and identify data on code violations, defensible space programs, and the need for community education to increase local resilience. Utility providers can assess their networks for vulnerabilities related to aboveground infrastructure, current fuel and weather conditions, and right-of-way clearance practices to reduce fire ignitions during fire weather. Each of these products can then be included in an integrated planning document such as a CWPP.

GIS in action

This first section presents real-life stories about how wildland fire organizations use GIS to map, analyze, and prioritize preparedness measures.

RANKING WILDFIRE RISKS TO PROTECT VALUABLE PLACES

US National Park Service

OR 95 YEARS, THE FERN LAKE BACKCOUNTRY PATROL Cabin inside Colorado's Rocky Mountain National Park sheltered rangers, biologists, and search and rescue crews. Then two wildfires imperiled the park in late 2020, and one—the East Troublesome Fire—made it through the park entrance, growing to more than 100,000 acres in one day.

The cabin didn't stand a chance against the blaze. Neither did the Grand Lake entrance station, the historic Onahu Lodge, the Green Mountain cabins, or the Trails and Tack Barn. The damage could have been much worse, though, if not for the work years earlier to remove trees that had been killed by pine beetles and could have served as fuel.

Many enduring and recognizable landmarks inside the US National Park System are vulnerable to wildfires made more destructive by climate change. Warmer winters, for example, have allowed pests to thrive and infest thousands of acres of trees—increasing potential kindling for fires across North America.

As part of an effort to fortify those landmarks and structures, a small team at the National Park Service has collected data and assessed 98 percent of park structures agencywide. The team created an interactive wildland fire risk assessment site using GIS technology. The site's GIS maps and dashboards show the threat to each landmark and structure and detail what can be done to achieve greater resilience. Each assessed structure, whether a historic inn or a park entrance sign, is ranked based on vulnerability and value—all visible on a map. With a clear ranking of needs, National Park Service leadership believes it can more strategically invest in efforts to reduce wildfire fuels, especially with an influx of infrastructure funding on the way.

Maintaining the past, preserving the future

The National Park System has long maintained a facility asset management database with information about each of its built structures. But the database included just a single field with a yes-or-no answer to the question, "Wildfire risk?" The database was never intended to detail fire risk with information such as what kind of fuels might be nearby, or what resources were available to firefighters for defending a structure from wildfire. No complete spatial dataset existed for park structures either. If that information existed, it was in a paper record or stored on a hard drive, not in a shared place where it could be considered in the context of the entire National Park System. The database also lacked an accessible historical record of past assessments made during a fire. Fires could threaten a group of structures year after year, and each time assessments would be repeated from scratch.

Assessments populate a dashboard, which offers a map view of priority structures and treatment plans.

In recent years, the National Park Service began cataloging about 40,000 structures across the country. Sarah Hartsburg, project lead for the National Park Service's wildland fire risk assessments, and Skip Edel, Fire GIS Program lead for the National Park Service, thought there was a better way to collect, store, and make data about the structures available for planning and wildfire response.

Hartsburg and Edel's assessment work paid off in April 2022 in New Mexico as the Hermit's Peak/Calf Canyon Fire crept closer to Pecos National Historical Park—home to some of the earliest Indigenous people. Assessments of the park's structures were uploaded into the National Interagency Fire Center's (NIFC) Structure Triage dataset, providing information to thousands of firefighters and responders.

Although the assessments don't include natural landmarks such as trees or rock formations, they offered the first comprehensive look at each park's wildfire risks, with assets ranging from visitor centers and parking structures to modest "comfort station" commodes.

New buildings in the park system are often built with natural threats in mind, but safeguarding the parks' many historic structures can be challenging. To retain historic character, many of these structures feature original or similar construction materials and are not necessarily fire-retardant. In response, the assessments outline ways to fortify the buildings by removing anything around the structures that might fuel a fire while maintaining their history.

"We don't want to change the structure, but we don't want it to burn down," Edel said.

Seeing where the risks exist

The collected data from the assessments appears in an online dashboard that includes a map—searchable by park, region, or nationally—with a list of highest-priority sites based on wildfire vulnerability and value.

The historic Old Faithful Inn in Yellowstone National Park ranks first overall, for example.

Because the data is visualized in a US map rather than in a static list, decision-makers can also see clusters of structures in need of preventive measures, making it more efficient to implement necessary repairs at one time.

Anyone in the National Park System can sign in to the site using a smartphone, tablet, or computer to view the data. Fire managers can edit and update the assessment database, including steps taken to harden structures.

Assessing thousands of National Park structures

Hartsburg and a team of contractors working alongside local park resources used ArcGIS Field Maps, taking pictures as they noted details such as a building's roofing material, any nearby firewood, and the presence of propane tanks and reliable water sources.

In 2016, for example, the team visited Isle Royale National Park on Lake Superior in Michigan, accessible only by boat or plane. On her visit to the Daisy Farm Campground area of the island, Hartsburg noted a heavy fuel load of shrub and timber within 300 feet of a cabin, but she also observed a reliable nearby water source, a fire-resistant roof, and no overhead hazards. Recommendations made after the site visit included removing dead or dying shrubs and trees within 100 feet of the building and installing a sprinkler system to better protect the structure. Included in an assessment report is a detailed estimate of the labor hours or equipment needed to create a defensible space around the structure to protect it from wildfire.

Once the in-person visits were completed, the team moved into a phase where high-priority locations are to be revisited every 5 to 10 years. Each location will be assessed to see what treatments have been or still need to be made to make it more resilient to wildfire threats.

Top five National Park Service structures at risk:

1. Old Faithful Inn, Yellowstone National Park
2. Canyon Lodge, Yellowstone National Park
3. Canyon Visitor Center, Yellowstone National Park
4. Old Faithful Snow Lodge, Yellowstone National Park
5. Château, Oregon Caves National Monument

To prioritize the structures, the National Park Service uses data collected during the risk assessment to generate a "likelihood of ignition" score and rating. The criteria is borrowed from the National Fire Protection Association's Firewise USA program. That score is used in conjunction with the structure's "asset priority index" value from the Facility Asset Management database, which uses a 100-point scale and is based on criteria related to how the structure contributes to the park: resource preservation (natural and cultural), visitor use, park operations, and its ability to be replaced. Finally, the wildfire hazard potential—based on a geospatial product produced by the USDA Forest Service and Rocky Mountain Research Station—indicates wildfire likelihood and intensity across the United States.

"We throw those things in the GIS blender and come up with a score that we then use for ranking treatment priority and plan across the entire park service where to allocate limited resources," Edel said. "We have the data, we have the process, and we have the way to then allocate that money in the most prioritized manner. This is a really solid way to make decisions."

A version of this story by Anthony Schultz titled "By Ranking Wildfire Risks, National Park Service Protects Valuable Features" originally appeared in the *Esri Blog* on July 14, 2022.

MAPS CAN HELP SAVE US FROM CLIMATE-FUELED WILDFIRES

CAL FIRE and Technosylva

I N MANY WAYS, A MAP TODAY SERVES THE SAME PURPOSE as civilization's first maps. It synthesizes where to go, how to get there, and the hazards to avoid but in smarter and more collaborative fashion.

We need maps not so much for finding our way through forests but to show us how to save them, and us.

Wildfires have destroyed millions of acres of California forest in recent years. Entire towns have burned. Some insurance companies have stopped writing new homeowner policies, citing the wildfire threat.

Modern mapping tools are beginning to help fire crews tamp down these combustible events made more unwieldy by climate change and guide practices to improve forest health.

CAL FIRE deploys mapping strategies from Esri® partner Technosylva, with a predictive capability (to see where fires are sparking and might spread) and a tactical application (to track in real time where trucks, dozers, field crews, helicopters, and other assets are at any moment on a shared map).

In the context of wildland fires, today's dynamic digital maps aim to safeguard lives and property.

Nonetheless, the burning of so much wildland in recent years exposed the root problem that mapping technology can also help solve. Forests haven't been allowed to naturally burn for more than a century. Loaded with dry or dead branches and trunks, the fuel has helped lead to megafires. Now, human intervention is needed to thin forests to take on the role that fire has played in the past.

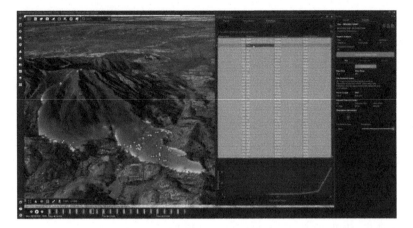

CAL FIRE uses Technosylva's Wildfire Analyst tool to understand the impacts of a fire's predicted spread. Dots represent structures on the map, and a corresponding table indicates when the fire may reach these buildings. With this detailed simulation, CAL FIRE can quickly understand the potential for the fire and make better decisions on response and suppression options.

Thinning of forests is one step toward a future with better forests that burn less and ultimately store more carbon so it doesn't warm the atmosphere. Accomplishing this goal requires maps to show how things should be and reveal what has been broken.

Entire landscapes are changing because of climate change, and trees take centuries to slowly march across the landscape on their own. Human intervention can expedite this process when possible. Modern maps are guiding that ecosystem transformation.

After the devastating 2018 Camp Fire burned in Butte County, a new climate-aware reforestation plan was enacted by the US Bureau of Land Management. Instead of planting back what burned, foresters planted trees a thousand feet higher in elevation, where they will stay cooler in soils where they can thrive. Rising temperatures are making it harder for many species to survive where they have lived for centuries. Maps depicting these changing conditions and the combination of factors necessary for healthy forests, such as soil characteristics, elevation, aspect, and precipitation forecasts, underpin forest restoration efforts.

In California, Governor Gavin Newsom secured major funding increases for active forest management, and in 2022, the US Forest Service released a 10-year strategy to confront the wildfire crisis with the $3 billion from the Bipartisan Infrastructure Law. Both plans acknowledge the need to improve forest health as part of climate action, with funding focused on preventing wildfires and restoring forests. Both plans require advanced mapping to inform current conditions, spatially allocate mitigative measures, and simulate expected outcomes.

Many collaborative statewide mapping efforts have been undertaken, including efforts by the California Natural Resources Agency (CNRA) in a public-private partnership with Esri. The shared mapping resources of these efforts bring together the knowledge of diverse experts to provide a sophisticated understanding of California's lands, forests, and waters. This information will guide forest transformation.

The tactical and analytical tools of modern maps can help us navigate the challenges of climate change as they relate to wildland fires and work to save our communities and forests.

A version of this story by Ryan Lanclos originally appeared in the *Esri Blog* on January 24, 2022.

EMBRACING APPS TO BETTER PREPARE FOR EMERGENCIES

Tucson Northwest Fire District

I N THE ARID DESERT LANDSCAPE OF SOUTHERN ARIZONA, where urban communities border wildlands, emergency preparedness has taken on a new urgency. Fires in Australia, Canada, and the United States, where thousands of homes have been lost to wildfires, remind firefighters and homeowners of the potential for catastrophic damages.

Firefighters at the Northwest Fire District (NWFD) in Tucson, Arizona, work to prevent fires and keep their community and its firefighters safe by inspecting hydrants and performing preincident planning. In addition, the NWFD team runs outreach programs and business inspections to help avoid injuries when fires start. NWFD ensures firefighters can locate the closest hydrant, be certain

Quick action by the Northwest Fire District put out a brush fire that threatened nearby structures in Tucson in 2019. Photo courtesy of Northwest Fire District.

it works, and determine whether hazardous materials are stored in the involved structure or any nearby buildings.

With GIS, the NWFD team recently created mobile field apps to speed its inspection process and data collection, achieving a new level of hydrant inspection efficiency.

NWFD has delivered a direct monetary benefit to the community from its improved preparedness. Insurance companies review many safety-related parameters when setting rates, including measurements of functional water supply, including hydrants. NWFD was given the highest classification, driving down fire insurance rates—saving money for everyone in the district.

"The evolution from a paper-based system to field apps has been very well received," said Jim Long, senior GIS analyst at NWFD. "Going digital means firefighters can easily collect, upload, and view field data on their mobile devices."

Creating confidence in hydrants

NWFD's 11 fire stations serve a growing community of more than 110,000 people within a territory that spans more than 150 arid square miles. With little surface water available, firefighters rely on the operational status of each hydrant. They need to know that the hydrant they will be using on scene is functional.

Now that NWFD has deployed mobile apps, every user can see the operational status of nearly 10,000 hydrants in real time.

"Our force can quickly and easily pinpoint the exact location of hydrants, verify they are fully functional, and identify the water companies responsible for each hydrant if they aren't working," Long said.

Firefighters inspect every hydrant in their region annually and record the operational status. In the past, they used paper forms and maps that resulted in more than 800 pages of handwritten notes held

Firefighters inspect every hydrant in the region every year and record the operational status on the mobile app. Photo courtesy of Northwest Fire District.

in large binders that firefighters carried around in their trucks during inspection season. Information on each hydrant was hard to update and was often inaccurate and inconsistent from truck to truck.

Using apps, the NWFD team inspected and uploaded the real-time status of 4,000 hydrants in just 60 days—a workflow that makes hydrants more reliable and inspections transparent. "Seeing which hydrants are in what stage of inspection or repair on smart maps is huge because it allows us to track the progress in real time," said Tom Krinke, firefighter and paramedic with NWFD.

Making infrastructure inspections digital

Success with the hydrant app spurred the team to create and try purpose-built applications for additional preparedness tasks. A recent pilot test focused on preincident planning and risk assessment workflows, testing a mobile field app called the Pre-incident Plan Locator.

During the test, firefighters used the app to tailor surveys built to record the multihazard and integrated risk assessment workflows that are currently manual, and paper based. The app recorded details about buildings, beyond the basic floor plans, which get updated on an annual basis to produce hard copy maps.

As firefighters walk through a building inspection, they use the app to pinpoint and mark physical assets, such as the location of electric, gas, and water shutoffs; stairwells; elevators; and the alarm control panel. They can also assign a risk assessment score to each address. Structures are then color-coded on smart maps, providing an instant visual capture of the risk level, which firefighters can quickly reference en route to a fire.

With the app, if the structure is coded for higher risk, firefighters can access more detailed data about the exact location of hazards before entering a building.

The prototype app could aid with such practical decisions as where to park the fire truck as it illuminates an 800-foot radius around a fire that corresponds to the length of the hoses on every truck. It also identifies viable hydrants for the incident.

In the dry region of the US Southwest, NWFD firefighters will continue to battle blazes, protecting people and property using every advantage they can muster—including technology. The team's recent creative use of apps and smart maps has the potential to further improve preparedness.

"It's a matter of time before we take this preincident plan system to the next level because fire captains only have 60 seconds to get all the information they need before each response," Long said.

A version of this story by Mike Cox titled "Northwest Fire District Embraces Apps to Better Prepare for Emergencies" originally appeared in the *Esri Blog* on July 14, 2022.

STUDENTS PROTECT THE UNHOUSED FROM WILDFIRES

Anderson W. Clark Magnet High School

KNOWING WHERE UNHOUSED POPULATIONS ARE sheltering is vital during wildfire evacuations. The numbers of homeless people have grown in California. The forests, fields, and hills in urban areas are full of dry brush because of drought. The results of the 2020 Greater Los Angeles Homeless Count showed that the number of residents experiencing homelessness grew 12.7 percent in 2020 to 66,426 people—many of them sheltering on the dry, open land. By 2022, the homeless population had increased to nearly 70,000 in the greater Los Angeles area.

The students from Anderson W. Clark Magnet High School, whose school borders the Angeles National Forest in Glendale, set out to prove that infrared images analyzed with GIS technology could identify homeless encampments near areas at high risk of wildfire.

Unhoused people often congregate in parks and open space and may not know about the wildfire hazard.

A geographic approach to research

The curriculum at Anderson W. Clark Magnet High School integrates technology with core academics, including an honors course in GIS remote sensing. Spanning six weeks in 2021, the class had 12 students collaborate to build, program, and test what's known as a cube satellite, or CubeSat, prototype. These square miniature research satellites are used to collect data and take measurements. CubeSat prototypes are built with inexpensive materials and tested at low layers of Earth's atmosphere using balloons, drones, amateur rockets, or small aircraft.

The students' CubeSat project was part of the US Department of Education's CTE Mission: CubeSat challenge, which calls for designing, building, and testing a cube satellite prototype to tackle issues important to local communities.

To make the prototype, the students chose infrared sensors that could detect heat signatures and a GPS receiver to map the locations

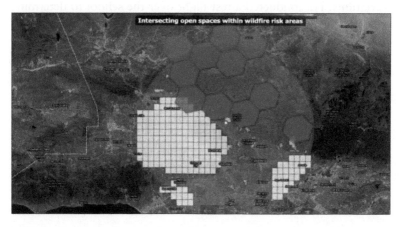

To find a test site, students performed a site suitability analysis within a 15-mile radius of their school that considered fire risk, land type, and Federal Aviation Administration restrictions that would dictate what types of test vehicles could be used. The areas in green on the map indicate the intersection of all factors.

of high-risk encampments. For the building phase, students were divided into two teams: one for the flight station and the other for the ground station.

Ensuring that each student had an opportunity to design a piece of the project, they self-assigned roles based on interest. Each student was also responsible for a secondary role, such as managing documentation, writing the flight report, or determining the best location for the test. They learned to assemble and program their CubeSat prototype with instructional videos from their teacher as well as information on YouTube and online forums.

"Throughout the project, I consistently was on forums looking up how to make stuff work. CubeSats are not a widely used product, but there was so much information about them online," said Matthew Keshishian, one of the students who worked on the project.

Keshishian added that programming the prototype's onboard computer was personally challenging because he created three code variations. He conducted a test flight over the school to determine

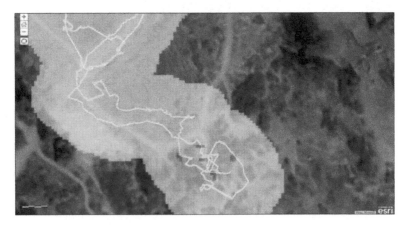

During the students' second flight at Hansen Dam, a small wildfire broke out, cutting short their data collection.

which code worked best with the sensors and camera. "We needed a code that was easy to use in most situations since we weren't sure how we would import the data into a map at that point," Keshishian said.

After building the prototype, the students searched for a test site. They performed a site suitability analysis using factors such as land type, locations without drone flight restrictions, and fire risk based on Los Angeles County data. In GIS, they layered the data to identify a high-risk area within a 15-mile radius of their school.

They chose the Hansen Dam Recreation Area as the best place to test their technology. Because of Federal Aviation Administration (FAA) airspace restrictions, students attached their prototype to a tethered weather balloon instead of a drone.

"A wildfire broke out at Hansen Dam halfway through our second flight and proved our fire-risk data was right," said Keshishian.

A positive impact on careers and community

The students shared their Hansen Dam Recreation Area web map and ArcGIS StoryMaps℠ story with Fire Captain Steven Marotta of the Los Angeles Fire Department. Marotta noted that dense vegetation makes evacuation efforts challenging and that the students' map would help emergency responders know where to search.

"This type of intelligence will help us to focus rescues and evacuations," Marotta said.

The team of students has since graduated, but their teacher Dominique Evans-Bye hopes to see the CTE Mission: CubeSat work continue in future class cohorts and to grow the project's impact. Evans-Bye said the project could one day include more geospatial technologies, drone imagery, and full-motion video captures through a partnership with the Los Angeles Sheriff's Department and Los Angeles Fire Department.

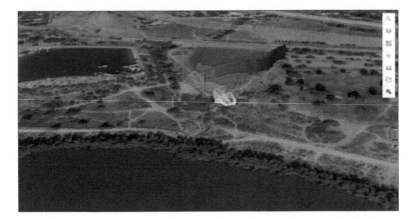

The Hansen Dam Recreation Area met most of the requirements for the GIS project.

Anderson W. Clark Magnet High School students received several accolades, including gold placement in the Industrial and Engineering Technology Cluster at the 55th annual SkillsUSA state conference and competition. Judges there noted the students' technical and workforce-ready skill sets, including hands-on engineering, programming, and mapping.

Evans-Bye agreed. "GIS impacts all industries," she said. "It is important to give students hands-on experiences like this where they can show off what they create while also impacting their community."

Knowing their work could succeed made their task more important. "I didn't just look at textbooks," said Keshishian. "I can look back at my high school journey and, from this project, say I participated in something that gives back to the community and can save lives."

A version of this story by Tom Baker originally appeared in the Winter 2023 issue of *ArcNews*.

FROM PAPER TO PIXELS: RETHINKING COMMUNITY WILDFIRE PROTECTION PLANS

Esri

WILDFIRES HAVE EMERGED AS ONE OF THE MOST COMMON natural disasters, posing a significant threat to public safety nationwide. Western US states have grappled with unprecedented fire seasons on a seemingly endless loop. Over the last two decades, numerous communities have embraced community wildfire protection plans, or CWPPs, to prepare for these incidents. These plans provide a comprehensive view of a community's wildfire risk, identify infrastructure, and show where mitigation efforts can reduce risk across the landscape.

The Infrastructure Investment and Jobs Act in 2021 allocated $1 billion to support the development of CWPPs and promote risk reduction. The approach to CWPPs dated to a time when technology was far from cutting-edge. Considering the obstacles faced by fire services in communicating wildfire risk and mitigation strategies to the public, it is evident that modernizing CWPPs is imperative for their future effectiveness.

The impacts of wildfire on human health, environments, and ecosystems

Today, more people than ever live in places where development meets nature, or the wildland-urban interface (WUI), where structures and other human development intermingle with vegetation. WUI is the single fastest-growing land-use type across the country, with one in three homes now located within WUI areas. When a fire occurs, the loss of homes, businesses, and other property can be catastrophic, leaving individuals displaced and communities having to recover from devastation.

The impact on the environment can be just as harmful. Although many ecosystems have evolved to depend on fire occurrence, when fires burn too hot and too severe, swaths of land can be destroyed, ruining habitat and natural resources.

Moreover, wildfires contribute to air pollution, releasing enormous amounts of carbon dioxide and other harmful emissions, such as fine particulate matter (PM2.5), into the atmosphere. These pollutants not only exacerbate climate change but also pose serious health risks to humans. One study estimates that in the contiguous United States, wildfire-related PM2.5 is annually responsible for 4,000 premature deaths with a corresponding economic loss of $36 billion.

The consequences of wildfires extend far beyond the immediate destruction. Damage can continue long after a fire has been extinguished and the cleanup process has ended. For instance, soil can become unstable after a wildfire has burned vegetation—trees, shrubs, and other ground cover. Once the soil in these areas becomes unstable, it is more susceptible to erosion from wind and water. Thus, the long-term risk of landslides and debris flows rises. This long-term risk underscores the need for effective wildfire prevention and mitigation.

The importance of accessibility and the planning process

CWPPs have traditionally been delivered as written documents, some more than 200 or 300 pages long. Plans can incorporate everything from models that depict the spread of fire and evacuation routes to the science and methodology behind vegetation type and health. Although these elements may be important to include, many fire management employees and most members of the public never engage with these materials. At times, much of this information is complex for fire management professionals, let alone the layperson or a community.

Once the fire-related information becomes accessible, anyone can peruse the plan online, engage with interactive risk maps, and explore fuel treatments and mitigation plans in 3D. This style of delivery helps pique interest and may even turn people into fledgling wildfire aficionados.

The digitization of CWPPs not only democratizes knowledge but also supports community collaboration in the planning process. By shifting the focus of CWPPs from merely serving to secure grants or collect facts and transforming them into a storytelling medium, the community and fire agencies build a sense of teamwork. A successful CWPP unites communities and stakeholders in the shared pursuit of identifying and reducing risks.

A changing landscape

During the past 20 years, the frequency and severity of wildfires have served to lengthen a fire season to year-round and produced record-setting destruction. For wildland firefighters, strategies and tactics have also evolved in response to the changing fire environment. Many organizations are turning to a digital tool called ArcGIS Hub for support.

ArcGIS Hub and why it's suited for CWPP development

ArcGIS Hub is a cloud-based engagement software as a service (SaaS) that enables organizations to communicate more effectively with their communities. You can create a hub site using Hub to aggregate resources and start conversations with internal and public audiences around a project, topic, or goal. Think of a hub site as a geoenabled website.

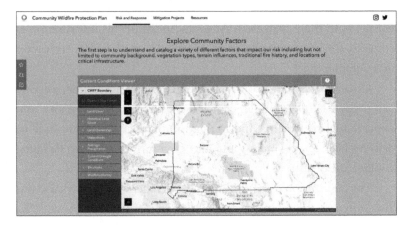

The first step of the planning process is to understand and catalog community factors related to wildfire risk.

The transition from paper to digital CWPPs offers benefits such as increased accessibility, real-time data updates, and a more interactive experience for community members. Hub allows stakeholders to create, manage, and share digital plans, enabling organizations and communities to stay connected and informed throughout the planning process.

Hub provides an intuitive interface, encouraging participation from anyone from the layperson to the wildfire expert. Interactive maps, visuals, and other multimedia content make it easier for community members to understand the complexities of wildfire protection planning and contribute their insights and ideas.

Hub simplifies and accelerates the planning process by providing tools for data collection, analysis, and visualization. Users can access and integrate a range of datasets, such as local land use, infrastructure, vegetation, and fire history, to create comprehensive CWPPs. Hub enables stakeholders to identify patterns, assess risks, and prioritize mitigation efforts with greater accuracy and efficiency. The sharing features in Hub support sharing updates, soliciting feedback, and

Hub offers opportunities for community involvement in the planning process.

coordinating efforts in the planning process. This level of transparency and information sharing builds trust and cooperation, leading to more successful CWPP outcomes.

Conclusion

By embracing digital transformation and harnessing the capabilities in Hub, users can create more accessible, engaging, and effective CWPPs that reflect community needs and concerns and use technology to build a safer, more resilient future in the face of growing wildfire risk.

A version of this story by Anthony Schultz originally appeared in the *Esri Industry Blog* on April 25, 2023.

PART 2

MITIGATION

MITIGATION PROGRAMS AIM TO REDUCE RISK THROUGH strategies that reduce the amount of brush, dead branches, and other combustible material around buildings and residences and prepare communities to defend against fire. ArcGIS allows fire managers to implement and track the progress of projects to reduce hazardous fuels, determine defensible space between homes and surrounding wildlands, prepare homes to resist fire, and conduct prescribed burns for healthier forests.

Fuel reduction

Reducing the amount of vegetation in a forest and around homes and businesses can help communities reduce their wildfire risk. ArcGIS tools help agencies define project boundaries, identify vegetation to be removed, monitor fuel-reduction progress, and quantify the effectiveness of their efforts after a wildfire begins. Using ArcGIS technology can help agencies implement fire plans with real-time mapping and allocation of resources as firefighters respond to reduce the impact of fire on communities and the environment.

Defensible space and home hardening

Creating areas where firefighters can safely defend homes and ensuring that those homes are prepared for the worst can reduce their risk during a wildfire. Using ArcGIS technology, firefighters and other agencies can use mobile field data collection apps to assess the condition of residences at risk, identify ways to increase defensible space, and record recommended home-hardening actions that increase a structure's resiliency to wildfire.

Prescribed fire

Fire supports the health of forests and ecosystems. Agencies can use GIS to determine where to conduct prescribed fires while considering fuel type, topography, wildlife habitat, and weather conditions. Fire managers can also predict how a prescribed fire will behave and spread under a variety of conditions and view the impact of the resultant smoke on population centers. With GIS, fire managers can set prescribed fires in areas and under the conditions where it will have the greatest benefit.

GIS in action

The rest of this section will look at real-life stories about how wildland fire organizations use GIS to plan and support mitigation activities.

AS THE WAYS WE FIGHT WILDFIRES CHANGE, SO WILL THE FORESTS THEMSELVES

Esri and Technosylva

W HAT WE'VE LEARNED ABOUT WILDFIRES—THEIR BEHAVIOR, fuels, chosen paths—has changed as dramatically as the technology used to study them. As a result, the way we fight them is changing, too.

After nearly seven million acres were scorched in California in 2020–2021, the state and country realized the need to fight fires differently, spending billions of dollars to prevent fires rather than suppress them. California set aside $1 billion in 2021 for fire efforts focused on clearing forests of potential fuel such as dry vegetation and plans to spend at least $200 million annually through 2027. Nationally, lawmakers have proposed billions to do the same across the country.

Californians and much of the country must learn to live with good fire to prevent devastating fires and adjust to a future where familiar forests may never grow back. It will take careful human intervention to foster a new kind of wilderness, according to experts at the Geography of Wildfire and Forest Resilience: Preparing for What's Next conference in 2021.

Modern technology is helping to build the healthy forests of the future by observing, even predicting, fire behavior and helping determine which species of tree may endure climate change and the threat of wildfire better than others.

A complex fire quilt

The many factors driving shifting wildfire behaviors include climate change, weather, local winds, water-stressed vegetation, fuels, topography, soils, and human impacts.

The complexities resemble a "fire quilt," said Joaquin Ramirez, founder of Technosylva, which developed the fiResponse software system, built with ArcGIS technology and used by states and fire agencies in predicting and monitoring active wildfires.

Society must learn to live with the type of fire that keeps forests in balance and resist the temptation to rebuild homes in the same locations of previous, recent, fires.

The majority of California's largest and most intense wildfires have occurred in recent years. Despite that, it could have been far worse, Ramirez said. There were likely countless more acres saved because of early detection and simulations through technology, he said.

Technology has been made even more necessary to study fires as their size and severity have grown.

"We can't ground truth a million-plus acres of land each year," said Libby Pansing, a forest and restoration scientist with American Forests, referring to the practice of checking assumptions in person. "We can't do this work on the ground by ourselves."

GIS has filled the gaps where landscapes can't be inspected in person. When restoring forests, the group must target its efforts to areas with the greatest ecological need and potential. Following the 2020 Creek Fire near Shaver Lake, California, which burned nearly 380,000 acres, the group used GIS to create a heat map to show distances to the nearest green trees needed for natural regrowth where the wildfire had burned. Another map layer showed a one-mile buffer from surrounding roads to determine where forestry workers could most easily access areas for replanting.

Seeing through smoke

California has become NASA's laboratory for testing new instruments and developing tools as it observes wildfires that have almost burned to the edges of its own Jet Propulsion Laboratory campus in Pasadena.

Dr. David Shimel, the agency's lead for carbon cycle and ecosystem programs, said radar technology has made it easier to observe topographic landscapes and forest structure, allowing them to distinguish the types of vegetation below the tree canopy. Lidar has led to 3D reconstructions of forests showing what an area looked like before and after a wildfire.

Thermal sensors can measure evapotranspiration and show whether vegetation is water stressed, risking worse fire situations. Positioned on the International Space Station, the ECOsystem Spaceborne Thermal Radiometer Experiment on Space Station (ECOSTRESS), shows the most stressed areas before a major fire and how that stress abates, or doesn't, as a burn scar recovers. Scientists can see through smoke and observe the energy of active fires.

NASA's tools provide input into the models that American Forests creates for a science-based approach to reforestation on burn scars. The nonprofit has developed a restoration plan for the site of the 2018 Camp Fire, California's deadliest fire that killed 85 people and left a huge swath of forest without trees. In the three years that followed, two more large fires burned nearby.

Austin Rempel, senior manager of American Forests, showed how each of the three burn scars met like puzzle pieces. Something unusual was happening, he said.

The fires had also managed to destroy the landscape in a way that trees weren't going to naturally return.

"If you lose all of your trees that can produce cones, you're in trouble," Rempel said, referring to the importance of the seeds

NASA's ECOSTRESS radar technology supports the fight against fires (shown at left) in the western United States. Image courtesy of NASA/JPL-Caltech.

housed inside those cones. "Without intervention, there will be no trees."

As a result, his group and others are building a forest that will be better equipped for what's coming, whether it's more fires and worsening climate conditions. Using GIS, the group determined that the seeds of trees grown in and around Redding, California, have adapted to higher temperatures and would be better suited for planting in the Camp Fire burn area as it's brought back to life.

Where fire is part of the forest

Experts in fire behavior, forest management, and disaster response hope some of the state and federal funding goes to increasing data collection at the local level to help make communities more resilient, doing more to reduce fuels in forests statewide, and additional research.

"We still don't have a good handle on predicting extreme fire behavior," said Craig Clements, director of the Wildfire Interdisciplinary Research Center.

Big fires, even intense fires, have been happening for a really long time, said Matt Jolly, a US Forest Service ecologist. What's different now is that fires are "more present in people's lives," forcing people to think about them in a different way. Technology can help provide the public context about why a fire is being allowed to burn in a controlled way to maintain environmental balance.

A healthy forest "is going to be a place where fire is part of that forest," Jolly said.

A version of this story by Kimberly Hartley titled "The Ways We Fight Wildfires Are Changing, So Will the Forests Themselves" originally appeared in the *Esri Blog* on December 13, 2021.

MAPS GUIDE WORK TO REDUCE WILDFIRE IMPACTS

Ashland Fire & Rescue

THE ALMEDA FIRE SWEPT THROUGH OREGON'S ROGUE Valley in September 2020, destroying or damaging more than 2,750 structures, including 2,400 homes in the region between the towns of Ashland, Talent, and Phoenix. The disaster reinforced a greater need for wildfire mitigation for each home and every neighborhood—a measure the City of Ashland is taking using smart maps and risk modeling.

Ashland Fire & Rescue's Wildfire Division Chief Chris Chambers recalled the devastation. "I drove right into the flames of an entire neighborhood burning. The initial priority had to be getting people out, not putting the fire out."

Jason Wegner, the city's GIS Manager at the time, assigned to a damage assessment team, was also haunted by the wreckage. "When I drove home that day, I thought to myself, 'Oh wow, that car driving toward me has paint on it.' It seemed strange since I'd been looking at so many cars burned to bare metal."

In Ashland, only a few structures were damaged, and during the fire many people were going about business as usual. As Wegner recalled, "There were people hanging out in the park and walking around with their lunches, with complete destruction happening mere miles away."

Although strong southeasterly wind spared the city from the devastation suffered by neighboring Talent and Phoenix, Ashland residents, firefighters, and city officials knew it was just a matter of luck, and the fire served as a wakeup call for the community. They emerged from the experience with added motivation to protect

This map shows the wildfire risk score of homes across Ashland, with red indicating highest risk.

themselves against the next inevitable wildfire—when the winds might not blow in their favor.

Learning from devastation

The verdant forest surroundings that coax people to call Ashland home also represent an ever-present wildfire risk. For residents, the Almeda Fire showed that firefighters may not have the time or ability to protect every home, that keeping the city and its people safe in the future will require a community effort, and that the actions of each homeowner affect the safety of their neighbors and entire neighborhoods.

Ashland Fire & Rescue had already been working on these issues for years, researching the public's understanding of wildfire risk, planning and coordinating mitigation efforts, and engaging homeowners in improving defensible space on their properties.

The Jackson County Fire Damage Assessment Dashboard details the properties damaged during the Almeda Fire. Clicking on the map reveals details and photos of the damage.

In 2018, fire department officials went house to house doing baseline curbside risk assessments of all single-family residences in Ashland. They performed the risk assessments using Intterra's firefighting software, powered by GIS.

Data from the curbside assessments combined with other data and analysis delivered a more accurate picture of risk. City of Ashland GIS professional Rickey Fite worked with fire department staff to build a model to calculate a risk level for each property, accounting for factors such as slope, aspect, road condition, types of vegetation and maturity, home construction, and vegetation proximity.

After the fire department used the model to assign each home a wildfire risk score, they communicated those details to residents to encourage home hardening and defensible space improvements around homes.

The citywide assessments directly contributed to the application and award of a $3 million FEMA Predisaster Mitigation grant. The funding is being used to assist with the creation of defensible space

Jackson County Fire Damage Assessment Dashboard			
Total	Residential	Commercial	Public
3,615	**3,255**	**348**	**11**
Properties Assessed	Properties Assessed	Properties Assessed	Properties Assessed
Total Destroyed	Residential	Commercial	Public
2,661	**2,483**	**173**	**4**
Properties	Total Destroyed	Total Destroyed	Total Destroyed
Total Properties	Residential	Commercial	Public
19	**11**	**6**	**2**
With Major Damage	With Major Damage	With Major Damage	With Major Damage
Total Properties	Total Residential	Total Commercial	Total Public
57	**52**	**5**	**0**
Affected	Affected	Affected	Affected
Total Unaffected	Total Unaffected	Total Unaffected	Total Unaffected
787	**635**	**148**	**4**
Properties	Residential Properties	Commercial Properties	Public Properties

The total counts of property damage across Jackson County from the Almeda and Obenchain Fires provide sobering statistics that motivated homeowners to make their homes safer against wildfires. The model accounted for other external variables as well, drawing on custom maps of all wildland vegetation within one mile of Ashland, building footprints, and a "nearby neighbor" analysis showing how a home's closeness to other homes could affect its chances of igniting. Whereas a typical calculation might apply the same risk score to a house whether it was located downtown or in the forest, the Ashland team developed this advanced model to better measure true risk exposure.

through vegetation removal and wood shake roof replacement for the top 1,100 at-risk home parcels to help reduce wildfire risks for everyone in Ashland.

Mapping defensible space

Individual homeowners can protect their homes against wildfire by creating defensible space—removing flammable surface vegetation within 100 feet of their home, eliminating ladder fuels that transfer fire from the ground to taller trees, and increasing spacing within the tree canopy. Special attention is given to the zero- to five-foot zone, which is the most critical to keep a house from igniting.

In Ashland, the work starts by removing flammable vegetation such as juniper bushes and making wood attachments to the house, such as decks and fences, more fire resistant. Material choices for roofs and siding make a big difference, as does hardening any areas where sparks could enter a home—for example, by installing 1/8-inch metal mesh screening on crawlspace and attic vents. During the Almeda Fire, many homes were spared initially only to burn down hours or even a day later because bark mulch that was placed against the house caught fire. New construction code in Ashland requires bark mulch to be at least five feet away from buildings.

Brian Hendrix, Fire Adapted Communities coordinator for Ashland Fire & Rescue, was involved in the initial data collection for the curbside inspections in Ashland and also assisted with damage assessments in Talent and Phoenix after the Almeda Fire. Drawing on those two experiences helped him understand how comparable hazards and risks in Talent and Phoenix affected structure survivability—what types of defensible space and home hardening techniques worked and what didn't. In many instances, homes that were lost had issues identified as wildfire hazards that increase structure ignition risks.

Hendrix says, "GIS has helped track hazards in Ashland, such as excess vegetation and combustible storage near homes, and seeing postfire remnants of those same hazards around the homes that burned confirms we've been giving residents the correct mitigation recommendations and wildfire risk warnings."

It's a virtuous cycle—assessing and mapping an area with GIS before a fire helps build data that can inform mitigation efforts, response, and recovery, which then informs future mitigation efforts. Fire departments can use a geographic approach, building a map of homes and vegetation that highlights risks as a foundational step to support action.

This map details fire risk scores in Ashland, giving residents a metric for improvement and an understanding of the fuels that wildfires feed on.

Nikki Hart-Brinkley, a GIS consultant serving the Rogue Valley, describes how a core set of data (vegetation health, building footprints, site addresses) is needed for all phases of a disaster. "If we're in resiliency building, if we're in recovery, if we're in response—it's critical to maintain these GIS resources so that we're ready to respond, and we're effective in our mitigation planning."

Prioritizing outreach and engagement with maps

Ashland Fire & Rescue has found the most effective way to encourage residents to harden their homes and improve defensible space has been through one-on-one engagement. When specific hazards on and around a home are identified and explained, the resident is more motivated to respond.

Ashland residents can request a personalized, in-depth wildfire risk assessment from the fire department, conducted largely by trained volunteers. Residents who have completed an assessment better understand wildfire hazard concepts along with specific mitigation needed.

Inspections can help residents be more mentally prepared for a fire by identifying areas of risk around their property and discussing ways to reduce those risks. Other topics covered during an assessment include receiving emergency alerts, evacuation protocols and "go bags," and encouraging communication between neighbors to enhance mitigation actions within areas that share topography and landscaping. Hendrix says, "The people we engage with appreciate the direct conversation and tend to feel less anxious after an assessment, which means they are less likely to panic during an event."

Ashland Fire & Rescue is also incorporating the use of GIS maps to track and share where risk assessments have happened in public maps, to help fire officials and concerned citizens in these communities identify areas where more outreach is needed.

Local knowledge builds better maps

One of the keys to Ashland's success is the level of detail and rigor applied to risk assessments and mitigation efforts. Rather than classify large swaths of the city at high or low risk, the fire department has used GIS to capture specific risk and hazard details at the parcel level on more than 6,800 residential properties through expert curbside assessments combined with exposure modeling.

Chambers explained the advantage this gives Ashland: "The wall-to-wall data we have is unusual with over 6,800 data collection points that help map wildfire risk to people's homes. We have a better picture of risk with site-specific data of fuels and homes and vegetation across a broad area within the city and the areas that surround it."

A version of this story by Mike Cox and Anthony Schultz titled "Ashland, Oregon: Maps Guide Work to Reduce Wildfire Impacts" originally appeared in the *Esri Blog* on December 15, 2022.

TARGETED GRAZING: HOW COWS PROTECT COMMUNITIES FROM WILDFIRE

Government of British Columbia, Canada

IN BRITISH COLUMBIA, CANADA, AN INNOVATIVE WILDFIRE mitigation program uses cows to forage in forests to reduce the intensity of fire.

"Cattle are already on these landscapes, and we're just using their grazing pressure in a different way," said Amanda Miller, an ecologist at Palouse Rangeland Consulting who was contracted to conduct the fieldwork for the program. "They're bulk grazers—better suited than goats or sheep to eat grass-dominated fine fuels that burn—and they don't need protection from predators in the same way, because a coyote can't take down a 1,200-pound cow."

The program started out as a pilot project, with GIS used to stratify landscapes, analyze the best test plots, and monitor grazing effectiveness. It's been deemed a success and will continue, because it's good for the forest and the cows.

Wildfires have always been part of the natural landscape of British Columbia. Occasional summer burns have long kept the scenic vistas of towering forests and stretches of grasslands in balance, removing dead ground cover and protecting trees from more intense fires that threaten the delicate ecosystem. In recent times, climate change and human management have altered this balance, causing larger and more frequent wildfires with significant consequences. Huge fires across Canada's 10 provinces and three territories in 2023 marked the worst wildfire season in Canadian history, created unhealthy air across much of the United States and Canada, and sent smoke around the globe. More than 34 million acres, an area the size of North Carolina, had burned across Canada by mid-August 2023, according to the Canadian Interagency Forest Fire Center.

"The intensity of wildfires in the province has been increasing over the past 10 years," said Shawna LaRade, the range officer with the Government of British Columbia (BC) who oversees the program. "We know the fires are not going away, and climate change is going to continue to influence the potential for catastrophic fires."

A partnership of scientists, cattle ranchers, and community

Researchers have shown targeted grazing can manage fire threat with minimal impact on the ecosystem. The grazing program is sponsored by the British Columbia Cattlemen's Association, after seeing how grazing changed fire behavior and acted to create "agricultural fire-breaks" during the province's devastating 2017 and 2018 fires. Participating ranchers are focusing the grazing effort on plants adjacent to communities in wildland-urban interface, or WUI, areas where summer-dried grasses accumulate and the potential for wildfire to threaten homes and communities is high. Grazing reduces fine fuels and promotes the growth of new, green grasses, which maintain moisture and burn more slowly. The newer growth is also shorter, and therefore less likely to spread flames to taller brush and trees.

Community response to the project has been supportive, with cooperation from forestry staff, First Nations partners, and city officials. "The City of Kelowna is a very engaged partner on one of our projects, and then also just the community members themselves," Miller said. "The local fire departments are super supportive of anything that can reduce the potential of structural fires."

The project, launched in 2019 with funding provided through the Canadian Agricultural Partners, the BC Ministry of Forests, and BC Wildfire Service, was inspired by the disastrous 2017 and 2018 wildfire seasons, which together burned more than 1,000 square miles of grasslands and forests. The goal was to discover whether targeted grazing could be used to mitigate the risk posed by wildfires to residents and local communities.

Ranchers hope to reduce the time they spend moving fences and managing their cows in the forest. Photo courtesy of Tyler Zhao, Columbia Basin Trust.

"Unfortunately, the grasses within the East Kootenay can quickly overgrow an area," said Mike Morrow, wildfire prevention officer for the SouthEast Fire Center. "While long and lush, the grass acts as a barrier. However, once that grass dries out, it can contribute to fire intensity and cause a significant increase in rates of wildfire spread."

A return to the natural cycle of wildfire management through grazing

In many of British Columbia's grasslands, wildfires are part of the natural cycle. The project seeks not to eliminate wildfires but to contain them and prevent the kind of catastrophic blazes that had burned more than 3,000 square miles of land in British Columbia during the 2021 season.

Particularly dangerous are crown fires, which burn hotter and faster between treetops, making them impossible to control and fight. "We feel the threat to Cranbrook is significantly reduced," Morrow said. "By utilizing cattle within specific areas adjacent to the private lands, if a fire were to start south of town and burn north, the fuel reduction treatment should lessen the intensity and help slow it down. Intensive grazing, coupled with landowners' own FireSmart activities, should help prevent fire damage."

Targeted grazing partially mimics the way British Columbia's grasslands have historically been kept under control, and herds of elk continue to roam and graze many of the project areas in the East Kootenays. Cattle are the most effective and efficient choice today, being plentiful and easy to manage. Additionally, grazing highlights the contribution that the agricultural industry can make to its communities with local cattle. "We're using existing resources; it's still part of the local food chain," Miller said. "I think it's positive all around, because it illustrates that we can support a locally sourced foodstuff while enhancing community protection values."

The allocation of forage to livestock is carefully managed to maintain healthy ecosystems, habitats, and a forage base for wildlife. "We have a significant wildlife population at the landscape level that we're managing, so a safe and allocated use for cattle grazing is an important part of resource management and sustainable, healthy ecosystems," LaRade said.

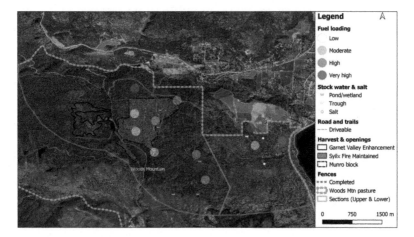

This map shows a gradient of fuel loading at the Peachland Targeted Grazing test plot in July 2021. Screenshot courtesy of the BC Government.

Monitoring fine fuel reduction with GIS

The researchers used field mapping tools and handheld devices to perform the initial and ongoing data collection on plant community cover, grass height, and changes in biomass as the grazing progressed. "We used GIS as basically a starting point for everything," Miller said. "Mapping is a heavy component of this, and quantifying the data visually through mapping products is a huge part of communicating our grazing strategy and outcomes."

GIS was used to provide maps for the ranchers to plan their next grazing locations. "We use ArcGIS with the most recent orthoimagery that we have," LaRade said. "Then we allocate forage production based on the vegetation types we observe within the treatment area."

During the last few seasons, targeted grazing on each test site reduced fine fuels by approximately 30 percent. "The forest fuels cleanup by cattle provides an extra level of protection," Morrow said. "We look forward to working with local ranchers and agencies on similar win-win projects."

The team has created heat maps to visualize the evolving fuel loads and track biomass reduction to demonstrate the program's effectiveness. "It's a really positive way of representing our work spatially, and it gets the message across," Miller said.

Researchers involved in the project have partnered with the Ministry of Agriculture to document fuel reduction. "We use maps to monitor progress on the plots," LaRade said. "We've got forest exclusion cages set up (to separate the ungrazed areas), and we're looking at forage production in the cages versus outside the cages."

The researchers are also monitoring grasslands and forest control environments to gauge outcomes of the program. "We have a lot of data to evaluate impacts to the ecosystem," Miller said.

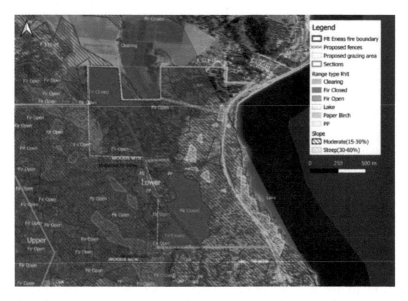

The pilot site map gathers details of vegetation type, slope, past fires, property boundaries, structures, and the proposed grazing area. Screenshot courtesy of the BC Government.

Greater understanding for future projects

In the WUI where the pilot project took place, mechanical harvesting was the first treatment to reduce the dense forests adjacent to the community. Grazing was the next treatment to reduce the grasses that arrived in the open spaces. In future years, prescribed fires will be conducted every 7 to 10 years to reduce the fuel load from any branches that have blown down. Fire will also rejuvenate the soil so grasses thrive and will eliminate younger trees, so the area remains open and fire resilient.

Program managers and scientists continue to fine-tune strategies to find the right amount of grazing for the best outcomes on the land. If grazing is too intense, it can lead to the introduction of invasive species, such as the cheatgrass that's taking over in other grasslands, including in Nevada.

Ranchers, too, are working to find a better system of moving their cows, because they currently have to erect and move temporary fencing, which is time-consuming and labor intensive. Virtual fencing, with collars on cows that can control their range and movement, is being tested in pilot projects elsewhere. This technological leap holds promise to manage cows through the use of maps and remove some of the burden and costs. It ties into a trend of GIS for operational intelligence, with real-time maps being used to improve the efficiency and safety of complex projects because everyone can see what's going on.

"Some of the tools that you can use for a greater understanding of what's happening are pretty amazing," LaRade said. "And people are accessing that information and creating this awareness that never existed before."

A version of this story by Anthony Schultz and Scott Noulis titled "Targeted Grazing: How Cows in Canada Protect Communities from Wildfire" originally appeared in the *Esri Blog* on November 1, 2022.

PART 3

RESPONSE

AGENCIES THAT PLAN FOR AND RESPOND TO WILDLAND fires learn more when they use tools that visualize and analyze data needed for fire suppression. Using ArcGIS technology, these agencies can visualize a fire's origin and its potential impacts to populations, the environment, and infrastructure. By managing resources with location-based technology, incident command staff can request, assign, manage, and track resources for a fire. Maps provide incident managers, elected officials, and the public information during a fire on activities ranging from response management to evacuating communities under immediate threat.

Visualize a fire's location

When a fire is reported, its location matters. By using GIS, fire staff can plot a fire's location and visualize its impact on populations, infrastructure, and environmental features. Learning who owns the affected land is also easily visualized and quantified. Fire managers can compare a reported fire's location to the footprints of past fires and identify natural barriers that can help determine the allocation of fire resources.

Respond

Response to wildland fires includes inventorying, managing, and deploying resources. With GIS, fire suppression and support assets can be queried based on resource type and location to find the closest resources available for a timely response. In the field, real-time mobile data collection capabilities help create a common operating picture for responders on an incident. As they share information in real time, agencies can track firefighter locations in relation to potential fire threats and fire movement.

Monitor

Effective response operations require situational awareness. An integrated, real-time, location-based solution helps incident commanders monitor changing conditions as they happen, brief staff and the public immediately in real time, and make better-informed decisions that can save lives and property. Location intelligence provides the mapping and analytics capabilities that allow fire personnel to see what and where things are happening the moment they occur.

Communicate

The public, residents in the fire area, and other stakeholders need access to information before, during, and after an incident. GIS can identify where roads are closed, where evacuation zones are located, and where to shelter. GIS supports an agency's public information program with visual data and storytelling tools. With GIS, agencies can report to the public with an authoritative voice using detailed, digital maps designed to scale when the public and media need them most.

GIS in action

This section will look at real-life stories about how wildland fire agencies use GIS for wildfire response mapping and support activities.

REAL-TIME WILDFIRE AWARENESS EMERGES FROM FIREFIGHTER CLOUD COLLABORATION

US National Interagency Fire Center

I N 2021, WILDFIRE BURNED THROUGH SEVEN MILLION ACRES in the United States—more than twice the annual average and higher than the six million acres that burned in 2022.

Indicative of this crisis's scope and impact on residents in 2021 was the record 2.2 billion visits to the National Interagency Fire Center (NIFC) open data site that shares the near real-time location and size of fires across the country.

"There are always the questions, 'How close is it to my house?' and 'Am I threatened?'" said Skip Edel, Fire GIS Program lead with the National Park Service who also administers the NIFC ArcGIS Online implementation. With access to fire perimeter updates on a map, residents can now discover the risk themselves.

What's known as the "NIFC Org" is the result of an ongoing effort to connect more than 22,000 users involved in wildland fire operations across agencies through a cloud-based GIS. It offers flexible reporting and visualization tools, providing opportunities to view and update operational details about most fires within the United States.

Before this digital transformation, with its mobile-first strategy, there were approximately 300 regular incident GIS users across agencies. By 2021, a complex mix of federal agencies, including the Department of Defense, 50 state and 6 territory emergency response organizations, members of the National Association of State Foresters, and hundreds of local responder organizations, each have access.

The National Incident Feature Service provides data about current wildfire incidents that can be added to a map, including on the Esri Wildfire Aware app that shows all active wildfires across the United States with important metrics about each fire. This screenshot was captured September 25, 2022.

"A large portion of the firefighters access NIFC Org using a smartphone or tablet," Edel said. "They love how they can draw in details on the map, make some notes, and then it shows up right away, helping them see their contribution to the overall effort."

Connecting with collaborators

Edel and the teams at NIFC work across member agencies, which include the Bureau of Land Management, National Park Service, US Fish and Wildlife Service, Bureau of Indian Affairs, and US Forest Service. Collectively, they manage wildland fire on nearly 700 million acres of federal public land and support the wildfire work in all other jurisdictions.

The NIFC Org hosts the National Incident Feature Service (NIFS), an interactive, authoritative data feed that further connects an already tight community that works long hours together every fire season. Some firefighters spend more than six months of the year moving from fire to fire. This makes the NIFS, with its easy-to-

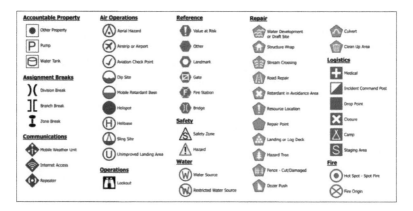

The National Wildfire Coordinating Group approved new symbology for the Incident Command System (ICS) standards for geospatial operations in April 2022. Many of the new symbols relate to distinct types of fire lines.

understand standardized symbols, a valuable reference point and collaboration tool for the on-the-move workforce.

"Standards help anyone who's working on the fire line understand what you're trying to communicate," Edel said.

Firefighters have been using standard symbols on maps for more than 50 years to depict such things as origin point, medical units, and incident command posts. NIFC created the NIFS feed in 2016, shifting to cloud-based ArcGIS Online so more users could access the information. Since then, the free flow of data has led to even more coordination among wildland firefighters across federal, regional, and local jurisdictions. The shared data provides the overall context of the incident, the current and planned actions of the crews, and the status of a fire.

Fine-tuning details to improve impacts

Internally, firefighters can use the NIFC Org to see details about their work assignments, such as drop points for supplies and rendezvous sites at the end of each day. Incident commanders look to it for an awareness of who is fighting which fires. News organizations and insurance companies have a more abbreviated view, with details about the status of fires through the NIFC open data site. For insurers, the awareness enhances risk abatement.

"Insurance companies are hiring their own engine crews to do their own firefighting," Edel said. "They might go out and prep a structure, take firewood away, cut down trees, put foam on the structure, and then either leave or stay and defend the structure."

For interagency wildfire operations personnel, seeing the real-time locations of fire resources—including air tankers, fire engines, and firefighters—on a live map is something that has long been envisioned and is now legislated.

Already, those fighting fires are seeing the benefits of mapping all available information. "When I show up on a fire in Idaho, everyone

A static map, such as this one in the crew camp of the Terwilliger Fire, still serves a purpose, but near real-time maps are quickly replacing printouts. The fire burned more than 8,000 acres of the Willamette National Forest in Oregon in August 2018.

knows the fire perimeter, because that data comes directly from the people who are doing the work," Edel said.

For now, when members of the public access NIFC open data, they can see only the fire perimeter. Soon, they will also see containment detail, which indicates where firefighters think the fire is no longer in danger of spreading.

Mobile-first workflows

The fact that the NIFC Org has scaled from 300 to more than 22,000 users within a five-year span speaks to the utility of the new digital workflows it makes possible. The old static map shared as a PDF and printed and posted on a big bulletin board at the fire operations center is still in use. But fire operations professionals are taking advantage of online mapping to decrease the time between collecting data on the fire line and GIS specialists providing an up-to-date map to inform critical decisions.

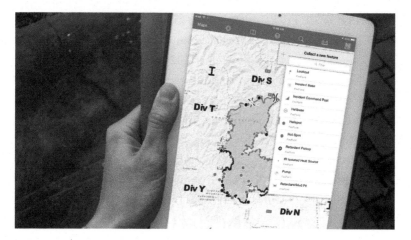

The Incident Web Map Template helps personnel on the fire line collect data that makes its way into many data products that span the full chain of command. Photo courtesy of the National Interagency Fire Center.

Fire personnel use such mobile apps as ArcGIS Field Maps and ArcGIS Survey123 to collect field data, even in areas without cellular service. Data and photos collected are then uploaded when cellular or Wi-Fi service is available.

The administrators of the NIFC Org have worked out detailed workflows and rules of behavior for those that use these tools. But the ease of making and sharing a map, with minimal training, is at the heart of its rapid adoption.

"The ability to put data and information together and get it out to people easily and quickly has been very useful," Edel said. "We don't restrict the public sharing of information during a fire. If you want to get a map out quickly to the public for an evacuation, for instance, we don't want any kind of process to hold you up."

A plan to add more fire history

In California, 2021 brought four of the 20 largest fires in state history, according to the California Department of Forestry and Fire Protection (CAL FIRE), compared to five in 2020.

California had not experienced another a fire in the top-20 category since then, as of mid-August 2023; nevertheless, federal and state funds have flowed in to address wildfires in California and across western states, with increasing budget allocations for fire response. Most of those dollars go to operational resources, such as fire engines, but there is also an imperative for more data to support better decisions.

Edel wants to create a single integrated, national dataset with incident history through the years showing who, what, and where details such as the incident location and the work of response teams. The idea is to fill in gaps about fire suppression efforts before modernization. For recent fires, he noted, "we can see every dozer line and everything we've recorded."

Fire data before 2016 exists in various formats. By bringing historical information into a digital space, on an interactive smart map, NIFC teams would have visualization, analytics, and an overall greater level of intelligence for mitigation plans. The research community could also use this data to dig deeper into trends and the cause and effect of fires.

"We still don't have a full view of landscape change through time," Edel said. "There's a lot of work to do to bring the data together."

A version of this story by Anthony Schultz originally appeared in the *Esri Blog* on August 2, 2022.

APPLYING NEW PREDICTIVE AND TACTICAL TOOLS TO FIGHT FEROCIOUS FIRES

CAL FIRE

C ALIFORNIA EXPERIENCED ONE OF THE MOST DESTRUCTIVE wildfire years in its history in 2021. The Caldor Fire became the nation's top priority in August as it threatened to jump into the highly populated Lake Tahoe Basin. The Dixie Fire from July to October destroyed the towns of Greenville and Canyondam and more than 1,000 homes, businesses, and other structures. In early October, the start of the strong Santa Ana wind season, the Alisal Fire near Santa Barbara spread quickly and temporarily closed Highway 101. Because of dryness and fuel loads, the fires generated their own weather, as thunderclouds massed above, causing lightning strikes that sparked spot fires ahead of advancing flames.

California Department of Forestry and Fire Protection (CAL FIRE) officials—who said they had never seen so many acres burn with such intensity—are using imagery, smart maps, and computer simulations to monitor and forecast fire behavior. "The technology helps us understand that without suppression, this is where the fire could go," said Phillip SeLegue, CAL FIRE's deputy chief of intelligence.

Allocating strained resources

Every day, CAL FIRE officials must continuously reprioritize resources for new and emerging fires. "Thankfully, we have a lot of resources here in California, but we also have a lot of ground to cover and people to protect," SeLegue said. Using a tool called FireGuard, CAL FIRE officials can see when and where fires spark, and then create strategies to match specific needs with strained resources.

FireGuard gives CAL FIRE access to video and images captured by classified US military drones and satellites as well as nonclassified imaging and weather satellites. Firefighters use the visualization tool on an hourly basis to monitor the fire's location, the heat it's generating, and the corresponding changes in weather conditions.

"The ability to immediately see that communities or houses are being impacted has brought more awareness to the situation," SeLegue said.

CAL FIRE combines FireGuard with Technosylva's Wildfire Analyst and Tactical Analyst products built on ArcGIS technology.

"Seeing at a moment's notice where all spot fire reports are coming in and coupling that with modeling tools from Technosylva lets us see where pinch points or problem areas are throughout the state," SeLegue said. Keeping on top of new and emerging fires helps

Each individual CAL FIRE user of Technosylva's Wildfire Analyst tool can configure their screen to display details of interest. Here, the analyst is looking at the current location of the fire and how it will likely spread, the real-time view from a nearby camera, details regarding the potential impacts and behavior of the fire, a list of recent fire spread predictions, and notifications of new fires that have been detected.

address the issue CAL FIRE faced in 2020 when several fires combined to create the August Complex Fire that burned more than one million acres.

Analysts with CAL FIRE can now run simulations, plan actions, and locate all firefighters, engines, helicopters, and other resources on a single shared map. In the field, firefighting crews can view this map on their mobile devices for alerts and fire activity awareness.

A bulldozer driver, for example, can see on a map where someone is scouting ahead and the locations, in real time, of arriving engines. "Instead of having to hail someone on the radio to ask their location, they can quickly see where they are. And if it reaches a critical stage, they can start building a safe refuge area that everyone in the vicinity can fit into," SeLegue said.

Embracing a shared map

The tactical and analytical tools from Technosylva add significant new capabilities beyond CAL FIRE's original mapping and analysis tools. California's 2021 fire season represented the first time an incident commander had a shared map with input and visibility from every crew.

"You can see your objective, capture your job, and see your edits as they go onto the map," SeLegue said. "The shared map provides a sense of accomplishment and encourages the use of this technology."

SeLegue's team has also found it helpful to have someone in the office, with access to other resources and analysis, updating the map with additional information such as weather and spot fires for those in the field.

"All edits are uploaded so others can see it visually and assist with the strategic plan or the overall goal," SeLegue said. "And I don't always have to start from zero, asking where all my resources are. It's just a click and I can see them."

Next, SeLegue hopes to integrate data, imagery, and videos collected from drones, which have become a flexible way for fire observers to monitor fire behavior without risking their lives.

Recovering from disaster

The new levels of information and automated workflows also support disaster recovery.

"We can use the data to create a Fire Management Assistance Grant, so communities can put in for federal loans and prop themselves back up and have better resilience," SeLegue said.

To SeLegue, the pain of fire damage became personal when the Carr Fire roared through his neighborhood and destroyed 14 of his neighbors' homes in July 2018. He lives in an area that is still recovering.

"These are cascading disasters," SeLegue said. The first is the fire itself. The second is the recovery—removing hazards, planting new trees, rerouting electrical lines and roads, and rebuilding. "As more fires occur, it continues to personally impact more folks in the fire service. It has pushed us to have a greater understanding of the effects of incidents. After the emergency is gone, what are you left with?"

The Dixie Fire and other recent blazes mark a pattern of rapidly expanding wildfires with erratic and unusual behavior, fueled by dry forests and severe drought conditions. The back-to-back blazes year-round leave little room to rest.

Firefighters like SeLegue are aware of the impacts on people's lives and the fact that firefighting conditions have become increasingly difficult around the world.

"The same issues that are happening in Spain or Perth or Queensland are the same issues we're dealing with here," SeLegue said. "I think the digital age has brought us together and made us more aware of other incidents and the solutions or tools we can draw on."

And SeLegue knows that being better prepared is not just limited to wildland fires. "CalFire is an all-hazards agency," he said. "We may be fighting fires today and responding to an earthquake or technological hazard tomorrow. These tools allow us the same level of situational awareness no matter what challenge we face."

A version of this story by Ryan Lanclos titled "CAL FIRE Applies New Predictive and Tactical Tools to Fight Ferocious Fires" originally appeared in the *Esri Blog* on October 18, 2021.

REAL-TIME TOOL TRANSFORMS WILDFIRE FIELD OPERATIONS

Pennsylvania Department of Conservation and Natural Resources

WHEN THE SEVEN PINES FIRE BROKE OUT IN NOVEMBER 2020, it quickly burned acres of scrub oak and mountain laurel along steep terrain near Swiftwater, Pennsylvania. Firefighters worked according to COVID-19 distancing needs and used live maps to concentrate containment efforts and protect homes. After four days, and with 812 acres burned, the human-caused blaze was out.

"This is 2020, so we're managing the fire while we're managing COVID-19," said Shawn Turner, forest fire specialist supervisor in the Pocono Mountains for the Pennsylvania Department of Conservation and Natural Resources (DCNR). "We couldn't have a whole lot of people on the fire, because then we'd have to deal with increased exposure."

The response effort included a regional team of firefighters and equipment such as a bulldozer to scrape the ground for fire breaks, a helicopter for targeted water drops, and reconnaissance aircraft to relay photos of the fire's extent. Firefighters mobilized this response with a live map app called the Fire Mapper app, built with GIS.

The Fire Mapper app let them see the work of others and report their own actions while keeping everyone away from each other because of the pandemic. It helped the teams orchestrate efforts as the fire and weather patterns changed moment by moment.

"We had very low humidity for two nights after the fire started," Turner said. "This was in a remote area with dry conditions that hadn't burned since the early 1970s, so the fuel load was high. We knew the fire could get large on us."

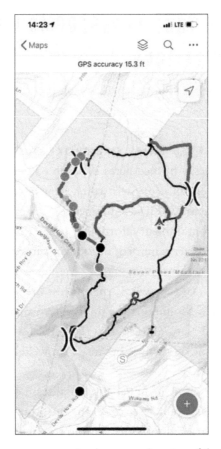

The live Fire Mapper app contains the current location of the fire line (in red) and the containment lines created by firefighters and bulldozers (in black).

Firefighters knew that if the fire crested a ridge beyond the state game lands where it originated, people in the 1,000 homes of the Pocono Farms East neighborhood would be forced to evacuate.

Instead, the humidity increased, and the fire kept advancing in a northern direction, staying below the ridge, and moving toward a more mature hardwood section of forest with fuels that don't burn as readily. "As soon as nighttime humidity recovered, like it normally does, it slowed the fire growth and gave us the time to get it," Turner said.

The live fire map takes hold

Turner, who functions as an incident commander, has been enthusiastic about using Fire Mapper to guide field operations.

"When I started in 1990, we still used a compass to do surveying for wildfires," Turner said. "We moved to GPS and GIS, but we still waited for a day or more to get a map out to the field. Now people are seeing what we can do with the live map."

The Fire Mapper app was first created during the 2014 fire season by Matt Reed, operations and planning section chief for DCNR Division of Forest Fire Protection, with help from Chad Northcraft, a fire forester turned air tanker base manager. The two shared a vision for a live map that could help teams efficiently fight fires. Reed built an app using ArcGIS Collector.

"There are four things that we're constantly aware of in fighting fire that we call LCES—lookouts, communications, escape routes, and safety zones," Reed said. "One of the first things we want to do is see where the fire is at and just pay attention. We always strive to have someone up in the air to relay information through the map."

Fire mapping was used selectively during the first year, but it caught on and was rolled out statewide in 2015. Anyone involved in wildfire or prescribed fire activities within DCNR has access to the app so they can record their work.

"You know the saying, 'A photo is worth a thousand words'? It is," Turner said. "When somebody's describing something to you, sometimes you're not exactly sure what it is they're talking about. But if they can take a photo and attach that to the map, then you get a really good idea of what's happening."

In 2016, dry conditions sparked 850 wildfires in Pennsylvania.

Firefighters start a burnout operation to remove unburnt fuels between the fire and containment lines during the Seven Pines Fire. Photo courtesy of Matt Reed.

In April of that year, the 16-Mile Fire burned 8,000 acres in the Pocono Mountains in northern Pennsylvania, requiring 130 firefighters who finally got it under control after two weeks.

"The 16-Mile Fire got very complex; it started as two fires that then came together," Reed said. "Fire Mapper got a lot of use and a lot of exposure because we had people from all over the state responding to that one and even people from outside the state."

Seeing changing wildfire behaviors

In 2020, five million acres across the western states of Washington, Oregon, and California were consumed by fires, including blazes that scorched more than 800,000 acres in California alone. The fires have been called climate fires because they reached farther and burned longer because of warmer and drier conditions.

Climate fires also threaten eastern US states, putting the 1.1 million acres of the Pinelands that span eastern Pennsylvania and

The new fire tower in Big Poconos State Park replaces a tower that had been staffed by fire spotters during the spring and fall fire seasons since 1921.

western New Jersey at high risk. Recent droughts and dead trees from spongy moth infestations have also heightened concerns for large fires in the Poconos.

Contrary to the Seven Pines Fire, it's not typical for a fire to burn for long or far in the Poconos in November. More than 80 percent of wildfires in Pennsylvania occur in March, April, and May. It's during those months when the winds are higher and the sun hits the forest floor and dries out the leaf litter that conditions are the worst for wildfires. In 2020, dry conditions created an unusually high number of fires in February—161 compared with 11 a year earlier. Scientists expect ongoing shifting seasonal patterns because of climate change.

The region's devastating 2016 fire season reinforced the need to invest in new fire lookout towers in the Poconos, replacing 16 towers built in the 1920s with new modern structures. Although most states are shutting down and reducing fire towers, Pennsylvania has found lookouts to be cost-effective and superior to other monitoring methods in remote areas, so it's building more.

"There are a lot of places where if people see smoke, they will get 10 calls to the 911 center," Mike Kern, chief of the DCNR Division of Forest Fire Protection, told National Public Radio. "There's other places in north-central Pennsylvania especially where no one will see a fire for a couple hours."

The towers are a front-line defense to spot smoke quickly and keep fires small. The live Fire Mapper app collects all inputs and displays data for everybody, including key inputs from lookout towers.

"I recently talked to one of our dispatchers who uses the tool in

The map contains the location of the fire line (in red) and the containment lines created by firefighters.

a district where towermen call in with the bearing and distance of smoke columns they see that she then pops on the map," Reed said. "Next, she sends the call out to see who she's got in the field. I'm working on a dispatcher dashboard so she can see where her personnel are, see where the fire is, and dispatch the closest person automatically, sending them a location via the map."

Five years of incremental improvements

Fire Mapper spans a broad number of users who each view the map and use it for their own needs while contributing to the shared awareness of others.

"We have a recon plane in the air anytime it's a fire day, and they're up there mapping fires," Northcraft said. "The more time I spend at the tanker base, the more I realize how the tool can be used in different ways."

Incident management teams use Fire Mapper to track the location of firefighting assets, look at vulnerable properties and structures, and figure out how to access a fire. Leadership also shares the information to keep the governor and the public informed.

"I was on the Seven Pines Fire, but my supervisor was not," Reed said. "He could just pull this right up on his screen and see what's going on and give his report to the state forester."

"It is important to have quick and accurate information when dealing with dynamic incidents such as wildfires," said Ellen Shultzabarger, Pennsylvania's state forester. "It's beneficial to have this nearly real-time data coming straight from the field for reporting and decision-making needs."

Before calling in resources, on-the-ground response teams put a virtual pin on each fire, get an idea of each fire's direction and how fast it's spreading, and draw a polygon around the fire on the map.

For the Seven Pines Fire, the response team achieved a new level of awareness because a larger group of responders recorded their actions using a new tracking function.

"I couldn't believe how simple it was to have the dozer boss mark his points," Reed said. "We were able to just convert the dozer line to the controlled fire edge. We had another guy scouting to guide the dozer. It was so simple to say, 'Hey, Greg, can you draw a line in where you're proposing that dozer line?' Greg says, 'Yeah, here you go.' And there it is on the map."

A version of this story by Mike Bialousz titled "Real-Time Tool Transforms Wildfire Field Operations in Pennsylvania" originally appeared in the *Esri Blog* on January 12, 2021.

VOLUNTEERS TRACK QUICKLY SPREADING WILDFIRES

#FireMappers

A S HE WATCHED WILDFIRES RAGE ACROSS CALIFORNIA, Paul Doherty had an idea. It was 2015, and Doherty was living in Redlands, 60 miles east of Los Angeles. He noted heightened levels of anxiety and alertness among people across the region.

The problem, Doherty thought, was how to take advantage of that alertness, especially in the initial stages of a fire. The Los Angeles metro area encompasses five counties and hundreds of independent cities. Could people turn to city government sources, county officials, sheriff's departments, or federal agencies for the wildfire information they need? Each has its own data and social media sources. Moreover, fires often travel across jurisdictions.

"If you didn't already know that a fire was occurring within the Riverside city limits, you didn't know to look for information from the city of Riverside," Doherty said. Evacuation information, often given only in text form rather than as a map, made it hard for the public to understand and act.

Doherty was proficient with GIS software used to create digital maps with place-specific data. GIS-based fire maps already existed, but they were mainly used by firefighters to track a fire's progress and only updated once a day. Every night, pilots flew over the area, gathering infrared data that was then processed by a GIS program to create the maps. The next night, they did it again.

Using his own ArcGIS Online account, Doherty began a project called #FireMappers. "I just started dropping points on a map where I knew there was a fire," he said. "And I knew because a reporter was there, or it was described on the radio or through social media—whatever I could figure out. I did it mostly for fires near Redlands, and mostly for friends, so they'd know what was happening."

Delineating the exact boundaries of fires in progress wasn't the point of #FireMappers. The goal was to give people a way to visualize their location in relation to the danger—and provide links to websites and social media accounts of responding agencies for more information.

Even after Doherty moved to Canada and then New Zealand, he kept the project going, monitoring internet feeds of Southern California police and fire scanners. #FireMappers developed a steady following.

A new kind of fire map

Doherty noticed that traffic to the online maps would increase when a fire encroached on a densely populated area, or when a fire seemed particularly complex and unpredictable. In October 2019, Angelenos experienced both simultaneously when the Getty Fire erupted near the I-405 freeway and burned through the hills of the Brentwood neighborhood. The presence of Santa Ana winds—dry blasts that blow down from the mountains, a fixture of autumn in Southern California—increased the fire's volatility.

Traffic on #FireMappers surged, from around a thousand visits a day to a million. "I was thinking, uh-oh, this is more success than I was looking for," Doherty said.

Doherty realized he could no longer treat #FireMappers as an experiment. By then, he was again living in Redlands, working as the director of technology innovation for the National Alliance for Public Safety GIS (NAPSG) Foundation. NAPSG agreed to bring #FireMappers under its umbrella, but the project would still need outside assistance.

The project expands

Doherty reached out to GISCorps, a volunteer offshoot of the Urban and Regional Information Systems Association (URISA), to tap the skills of advanced GIS users who design mapping projects for humanitarian or disaster-relief purposes. GISCorps quickly assembled a team of 30 #FireMappers volunteers, divided into a few geographic regions. Each regional group established a Slack channel to discuss information on emerging fires.

The addition of GISCorps allowed #FireMappers to expand its purview into Northern California and the Pacific Northwest, just in time for the latter region to experience a brutal wildfire season in 2020.

"We had a huge wind event and very dry weather, and so the sparks just took off," said German Whitley, an Oregon resident and GISCorps' project lead for #FireMappers.

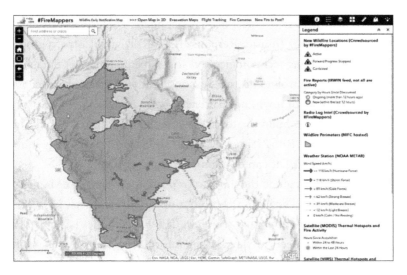

For the Mullen Fire in Colorado and Wyoming in 2020, #FireMappers used ArcGIS Explorer to add greater details about fire behavior.

The fire communication infrastructure in the Pacific Northwest was shakier than in California, where years of wildfires had instilled a homegrown situational awareness. "A lot of the fires here were in small and not very wealthy counties, where they watched their canyons burn," Whitley explained. "In Oregon and Washington, there's such a mosaic of jurisdictions—tribal lands, US Forest Service and Bureau of Land Management areas, state forests, and county sheriffs—and early in an incident there is very little sharing of information among agencies."

Tracking the information flow

Although #FireMappers is a mapping project, it's also a workflow experiment—an attempt to deal with large amounts of disparate information. In this sense, it mirrors recent developments in GIS, to optimize and organize geographically specific, near real-time data for clear communication.

To help with that aspect of #FireMappers, Doherty approached Crowd Emergency Disaster Response (CEDR) Digital Corps, a group that uses emerging technologies and social media to gather data and drive messaging around disaster response. "If you're the social media person for a small town, no matter how good you are, you're probably not going to build the town's Twitter account into a significant following," said Rob Neppell, CEDR's director of technology innovation. "What CEDR can do is amplify that message."

CEDR helps guide the flurry of reports reaching #FireMappers' Slack channels. Through a combination of curated lists, keyword monitoring, and automated searches, CEDR's guidance ensures that GISCorps' volunteers are getting the information they need as soon as it becomes available.

"There's the geographic space of fires—where they are," Doherty said. "And then there's this weird digital space—where's the info?

Without CEDR, we'd be good at mapping [fires], but not so good at mapping that digital space."

A version of this story by Mike Cox titled "#FireMappers Volunteers Keep Tabs on Quickly Spreading Wildfires" originally appeared in the *Esri Blog* on October 20, 2020.

REVAMPING WILDFIRE RESILIENCE AFTER DEVASTATING FIRES

Portugal National Authority for Emergency and Civil Protection

T ERRITORY MAPS COVER A WALL IN ALEXANDRE PENHA'S office, providing a constant reminder of how much has changed since he became operations deputy at Portugal's National Authority for Emergency and Civil Protection (ANEPC).

The country experienced its deadliest wildfires to date 17 days after Penha took over the post in 2017. Drought, high temperatures, and strong winds combined to spread 156 fires that June. Blazes burned 100,000 hectares, killed 66 people, and injured 200. Fast-moving flames destroyed enough crops and livestock to impact the country's economy. Many people lost everything.

In response, Portugal passed laws to reduce risk and better respond to emergencies. Penha's organization, ANEPC, redesigned its territories, adding five new regional operational commands. The added areas are designed to improve response times and the capacity to cope with increasing risks. ANEPC hired more firefighters, added aerial assets (helicopters and airplanes), purchased fire engines, and equipped firefighters with more protective equipment.

Penha looks at the wall maps to remember redrawn district lines that took effect last year: "There were 18 district commands and now there are 24 subregions, and each has a new name. It's not an easy test to remember each one yet."

These are some of the only paper maps he looks at these days because ANEPC has undergone a digital transformation. Penha now uses advanced wildfire mapping technology, including an early warning system aided by satellite and radar data to quickly detect wildfires. Real-time maps powered by GIS provide a live operational view

Alexandre Penha, operations deputy at Portugal's ANEPC, sits at his desk with maps of new regional operational commands on the wall behind him.

of every fire. Drones and helicopters populate maps with high-resolution images for up-to-the-minute awareness of fire locations and behavior.

Sharing data, seeing the same map

After the 2017 fires, upgrading the decision support system was the top priority. "At that time, information flowed in a text system, and someone was always trying to put that information on a map," Penha said.

Now a shared map captures incidents and activity so regional, subregional, and local stakeholders can see current conditions. The civil protection authority, the National Guard, the Forestry Institute, local authorities, and more, provide input and gain awareness through ArcGIS Online. Firefighters and the National Guard collect data on smartphones using ArcGIS Survey123. Helicopter crews use

The QuickCapture app helps helicopter teams quickly relay intelligence about wildfire behavior.

ArcGIS QuickCapture to gather images and make updates about fire behavior.

"They can send important information to the fire analyst—the type and inclination of the smoke, the wind, the type of fuel," Penha said. "The results were huge, not only for the flow of information but also our ability to predict the evolution of each fire."

The system provides a national overview along with a view of each subregion. It reports the events of the last few days, shows where fires are occurring, and tallies the energy released by each fire.

"Right now, on my screen, I have one fire with crews that have been dispatched, one where crews are working, one that has been contained, and one we're watching," Penha said. "I can keep an eye on each one. I can see how many men are there, how many trucks, how many aerial assets."

The map view integrates with other operational platforms, including weather predictions and dispatch centers. When a fire crew is sent to help, they know what to expect, and they can use

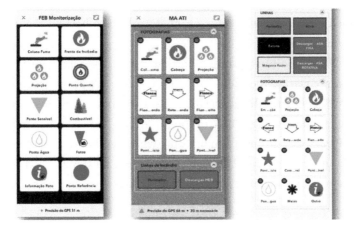

The QuickCapture app has large buttons and intuitive workflows to make field data capture easy.

the system to report their work. Later, administrative and financial details are added so all aspects of response can be assessed after each fire.

Reducing risk by reducing fuels

One of the first laws to pass after the 2017 fires was a mandate that farmers must clear their fields after harvest. The National Guard maps each farmer's preparations for the fire season.

"National Guard teams go from field to field to collect data," Penha said. "On the one hand, this work compels farmers to clean their fields. It also gives us an overview of what vegetation and fuels are present, so we know what to expect if fires occur."

In forests, prescribed burns reduce fuel loads. The mapping system identifies areas that should be burned and, in real time, supports firefighters in keeping each burn under control.

Penha was a firefighter in the Lisbon Fire Brigade for 20 years before coordinating a national response. With his new job, he has gained a national awareness of what's needed—beyond the use of high-tech maps.

"We have a huge problem in that the people who care for the forest are getting too old," Penha said. "Younger people want a different life. Even if the government has the money, if there's nobody there, nobody will take care of the forest."

Penha sees this problem not only with fires, but also with floods. Without people in rural areas to keep an eye on things, the pace of climate change overwhelms land managers.

Experiencing the worst and helping others deal with it

Regions with a Mediterranean climate are among the first threatened by growing wildfires. Portugal collaborates closely with other European countries, especially Spain, France, Italy, and Greece, as each has suffered major fires. But the threat is spreading.

In 2023, Italy deployed aerial assets to Germany and Romania because they are starting to suffer from climate change and have big fires, Penha said. "In 2019, we sent planes to Sweden, one of the countries that we never expected to have problems with forest fires. That's the way things are now."

In March 2023, ANEPC sent a team to Chile as the country fought several big fires during its summer months. Portugal's firefighters arrived equipped with the command-and-control system on smartphones. The 140 firefighters used QuickCapture to collect images and share information.

Drought and winds continue to plague forests in South America, as they do in Europe. In Portugal, the summer of 2022 was difficult. Fires burned July through October, inflicting widespread damage. Hundreds of people were injured, and hundreds of structures were destroyed. The military was called in to help with firefighting.

In preparation for its summer months, ANEPC developed AI

workflows to help determine where best to apply its resources. Teams were anticipating a need to battle between 80 and 120 forest fires every day.

"The number of fires will certainly not be fewer this year, and we won't have more firefighters," Penha said in early 2023. "We'll have some new sensors on our reconnaissance airplanes that will give us more information, and we have to do a rapid analysis of it. We're always moving volunteers and firefighters from the coast to the interior of the country and from the south to the north. If we don't have timely evidence, the places in need won't have firefighters."

A version of this story by Anthony Schultz titled "Portugal Revamps Wildfire Resilience after Devastating Fires" originally appeared in the *Esri Blog* on May 2, 2023.

PART 4

RECOVERY AND REHABILITATION

A FTER A WILDFIRE, RESPONDERS SUCH AS FIREFIGHTERS and utilities must assess damage and stabilize the environment before reopening communities. Landscapes will also need to be restored to ensure long-term recovery. Using GIS, agencies can quickly assess damage and remove debris allowing residents and others access to the impacted area. Once a wildfire burns through a community, the threat is not over. Several negative impacts from wildfire can still occur. By conducting a burn severity analysis, firefighters can identify areas at the highest risk of postfire flooding and landslides for emergency stabilization efforts. Agencies can also apply analysis to landscapes to determine action that restores environments over the long-term.

The importance of postfire recovery efforts and documentation of fire impacts continues to increase, as illustrated by the devastating wildfires in Canada, the United States, Australia, and other countries around the world. ArcGIS technology offers solutions that include imagery, machine learning, and AI to help automate damage assessment workflows. Additionally, emergency stabilization actions targeting areas identified as prone to postfire flooding, debris flows, and landslides can help mitigate postfire risk.

ArcGIS apps can be used to develop data for analyzing the effectiveness of current wildland-urban interface, or WUI, building codes and fire safety regulations. Mobile apps and analytical capabilities make it possible to analyze these standards with the use of verified data.

Damage inspection and debris removal

After a wildfire, it's important for firefighters, local government leaders, utility company staff, and others to assess the damage. Responders working in the field use mobile devices to collect data and pictures of destroyed residences, other infrastructure, and associated debris. This data can be mapped, inventoried, and communicated in real time to command personnel, who can plan for debris removal and expedite the process of reopening communities.

Identify areas for stabilization

Burn severity is a qualitative assessment of how much heat a wildland fire directed toward the ground. The most intense fires sterilize the soil, preventing growth for up to 10 years while significantly increasing risk of erosion. Burned areas can lead to postfire flooding, debris flows, and even catastrophic failure in the form of landslides. Using GIS and imagery, GIS specialists can determine the burn severity across a wildfire. Areas that burned most intensely can be targeted with postfire rehabilitation measures and stabilization efforts to prevent further impacts to the forest floor.

Long-term recovery and restoration

The need for rehabilitation following a wildfire does not diminish immediately after the fire is contained and emergency stabilization efforts are under way. It can take decades for burned areas to recover and soil to become productive. When fire managers map and monitor soil and vegetation changes over time using GIS, they can develop and modify recovery plans and manage these efforts long term.

GIS in action

This section will look at real-life stories about how wildland fire organizations use GIS for wildfire recovery and rehabilitation.

MAPPING A CLIMATE-RESILIENT FIRE RECOVERY PLAN

US Bureau of Land Management and American Forests

I N NOVEMBER 2018, AN ENORMOUS CALIFORNIA WILDFIRE claimed 85 lives and consumed the entire town of Paradise. Ever since, experts have been devising ways to safeguard against another tragedy and rebuild the forest destroyed by the Camp Fire.

Rather than simply replant what was there, the Bureau of Land Management (BLM) set out to map a climate-informed restoration plan.

"We want to plant it back better to withstand wildfire and future climate, so the community is not vulnerable like that again," said Coreen Francis, California and Nevada state forester at the BLM.

During her more than 20-year forestry career, Francis has seen shifts in forest health from drought, insects, disease, and climate. The pace of change in the forests around Paradise, however, has forced everyone to reexamine their understanding and try to catch up. To create a smart restoration plan, she convened experts to combine their knowledge about the land and forest using GIS technology to build a sustainable plan.

Several fires had burned across the same 153,336-acre Camp Fire burn area in less than a decade. Since then, more megafires have hit, including the North Complex fire that consumed 318,935 acres in 2020 and the Dixie Fire that burned 963,309 acres in 2021. Together, these fires have left few trees untouched in this corner of Northern California.

"This place doesn't want to be the same forest because it's so climate challenged," said Austin Rempel, senior manager of reforestation at the nonprofit American Forests. "For instance, sugar pine is

Wolfy Rougle of the Butte County Resource Conservation District surveys plans. Image courtesy of American Forests.

everyone's favorite tree because they grow big and look nice, but climate models say they don't want to live here anymore. Low-elevation sugar pine is going to be a thing of the past."

Assisting tree migration

Trees can't just pick up their roots and move, and a natural migration could take centuries. It's up to foresters to plant for what the forest wants to become, a practice known as "assisted migration."

"Assisted migration is a no-brainer for our organization, knowing that forests need to adapt," Rempel said. "In the Camp Fire area, because of its low elevation, it's quickly turning from dense mixed conifer forest into a place that wants to be more oak and grassland and chaparral and gray pine."

Analysts at American Forests apply models that use spatial analytics to consider species tolerances and soil types, along with climate forecasts on heat and rainfall, to predict what plants will survive in a place far into the future.

This level of climate action requires a detailed map to understand what exists, the conditions best suited for each plant, and where similar conditions can be found elsewhere. GIS is used to perform this suitability analysis with predictions that improve with more data.

For the BLM's Francis and other foresters, ArcGIS Online became the place to combine data and plan collaboratively. It also provided the portability for all to look at a shared map as they

The low-elevation foothills in and around the Camp Fire burn scar are warming up and drying out faster than any part of the state, aside from the Mojave Desert. Oaks are coming back because they have adapted to these conditions. Image courtesy of American Forests.

roamed the burn scar. "We take scientific concepts, and we look at them on the ground, and then we compare them with what we see on the map," she said. "We can scroll and look at different layers while we're walking to inform us of things that we can't readily see. Knowing the soil type is serpentine, for example, explains why those trees look scrawnier."

Getting the ground truth by checking the map in the field provides the opportunity to adjust and add more details.

"Some data we had was wrong," Francis said. "Being able to see it right there allows us to build knowledge and make our plan a little more accurate."

Planting and planning simultaneously

In California, the federal agency manages 15 million acres, much of which is inaccessible to crews replanting trees. There are also a limited number of seedlings, so they must be planted carefully where they will thrive.

"Based on capacity, resources, and access, we can only hope to reforest about 10 percent of the Camp Fire burn scar, and that's if everyone is working together," Rempel said. "That's another place where GIS comes in handy, because we have to be extremely strategic and know we're doing the right things in the right places."

Many of BLM's management practices are guided by shared maps. GIS is well suited to landscape-level planning because it contains details on the topography—ridges, rock outcroppings, slopes, water, valleys. Foresters must consider that north-facing slopes are cooler, south-facing slopes are drier, and valley bottoms have the deepest and best soils.

"Mapping the landscape is a starting point," Rempel said. "It shows us what the forest should look like and what we should plant there."

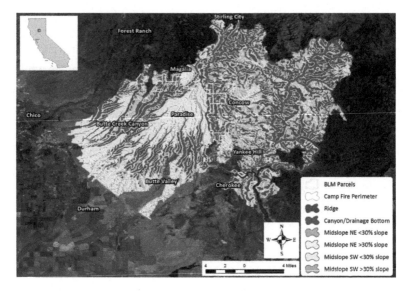

The topography—slopes and lowlands—determines most of the postfire forest management strategies.

The map pinpoints the places that will be climate stable and ideal for planting specific species.

"We know where trees live now, and we can model the climates they're comfortable with," Rempel said. "We can use GIS to map the soil productivity and where trees would be most successful."

The model and map include ecology, with data to analyze and explore the pieces of the environment that contribute to a tree's survival. GIS becomes a repository of earth processes and a way to query and model to apply nature-based solutions to restore a balance.

"We've talked about the concept of island plantings, where you put a diversity of species into a small plot, maybe a quarter of an acre, and grow those in clumps or islands across the landscape," Francis said. "Eventually, trees will produce seed and the seed will burst into the surrounding area, and it promotes more diversity on the landscape."

GIS also was used to plan and create natural fire breaks in the landscape to reduce the intensity of future fires.

The map helped speed the reforestation by picking the areas to plant first where they will have the most strategic advantage.

Commons for collaboration

Stakeholders and participants created the climate-informed restoration plan. BLM guided the effort with the help of American Forests and participation from the US Forest Service, CAL FIRE, Plumas National Forest, Butte County Fire Safe Council, Sierra Pacific Industries, and others.

Having a timber company at the table is unusual, but so is what happened to Sierra Pacific Industries' part of the forest that burned in 2012. The company diligently replanted it in hopes of harvesting lumber, and then six years later, the Camp Fire burned everything they planted. "That was enough for them to say, 'This is not a place where we can do production forestry anymore,'" Rempel said.

The stakeholders came to the planning sessions with ideas, maps, and open minds. The evidence was clear: everyone was wasting their time by doing the same things that had been tried before.

"Permaculture ideas—nature-based approaches—are starting to enter into forestry," Rempel said. "It takes a very long time to convince old-school foresters that this is the way, but it is happening slowly."

ArcGIS Online became the place where everyone could work and iterate together. For those not familiar with GIS, they could view the maps and agree or disagree with what they were presented.

"The sharing platform was central to our collaborative approach and our climate conversations," Rempel said. "We had these sessions during different versions of the draft where we got all the land managers and foresters together to go over what they were seeing or if other tricks of the trade should be added to the report."

Tackling a trend

The foresters who crafted the Camp Fire restoration plan hope that climate-informed strategies become more common. The approach is practical in making the most of limited resources by pinpointing the places where the forest can thrive.

"Many of the climate plans just offer big-picture ideas—about techniques that could be applied," Francis said. "Our plan takes those large concepts to the ground level. Predictions of what the climate is going to be informs our implementation plan."

According to research at American Forests, 81 percent of reforestation needed on national forest land is now due to wildfires rather than logging.

To replant wisely, new models must factor in future climate.

"This is a recovery plan," Francis said. "It's about using the best science to replant."

A version of this story by Mike Bialousz titled "Wildfire Restoration: Mapping a Climate-Resilient Camp Fire Recovery Plan" originally appeared in the *Esri Blog* on November 30, 2021.

USING GIS TO HELP CLEAN UP DEBRIS AFTER NATURAL DISASTERS

CDR Maguire

E IGHT WILDFIRES CONVERGED IN OREGON IN THE FALL OF 2020, and after the fires eventually quieted, the carefully choreographed cleanup effort began.

That's when engineering and disaster recovery firm CDR Maguire arrived to clear roads and rights-of-way for the Oregon Department of Transportation, returning some normalcy back to rattled communities. This time, though, there was a new challenge. On an Oregon cliffside, crews needed help removing charred, dead trees that risked falling onto the roadway below. Because the area proved hazardous for crews to do their jobs on the ground, they did their work dangling from helicopters, chainsaws in hand, They had done similar work before, guided by an arborist with binoculars to scout the site to ensure that trees that needed to stay, did so. But the vantage point wasn't ideal, and the crews needed to know exactly where and what not to cut.

"The assessment needed to be very accurate," said Deepali Datre, GIS manager for CDR Maguire.

There were plenty of unknowns, along with just as many new ways of approaching the challenge, she said.

Using drone technology, the company collected data to help officials better assess the hazards. Lidar sensors and photogrammetry techniques provided accurate images and 3D models that could be measured and analyzed to assess the hazards. The assist from the drone's vantage point gave the on-site crews a strategic game plan, including where to start.

Taking the gathered data and analyzing it in GIS, Datre's team

Drones have proven useful to assess the volume of debris piles left after a wildfire.

made maps and conducted spatial analysis to guide the cutting. An arborist on the ground, keeping their distance from the cliffside, still used binoculars to verify the trees that needed to go, but decisions were backed by shared detailed maps and 3D models to enhance the efficiency of the crews.

Cleaning up debris and keeping clear records

Debris removal operations can vary as widely as the natural disasters themselves, with conditions and circumstances driving technology needs. Drone imagery, for one, isn't always a necessity or used like it was in Oregon. Where trees are removed, CDR will occasionally be tasked with replanting new ones.

One aspect of debris removal is always the same, though. The work, often contracted out by local and state governments to experienced crews such as CDR Maguire, requires fastidious recordkeeping.

In Oregon, the cleanup involved the removal of 91,082 dead or dying fire-damaged trees that posed an ongoing hazard to roadways.

Ultimately, 158,689 tons of ash and debris were collected, clearing 3,012 private properties.

Debris piles are assessed, right-of-way permission to enter private lands is sought, and truckloads are weighed and then verified at drop-off. Every stop along the journey of the debris is noted and recorded using GIS. As communities recover and rebuild, they need the carefully kept records so they can be reimbursed for the costs—often millions of dollars—from the Federal Emergency Management Agency.

Following a fire in Colorado and later a tornado in Texas, the engineering company lost access to its usual vendor for recordkeeping and needed a new solution, fast.

Working with Esri partner Blue Raster, CDR Maguire customized Esri's Emergency Debris Management solution by creating a way for all parts of the complex data collection effort to function offline as there's often little power and internet connectivity in areas struck by natural disaster.

Many moving pieces, one custom fix

"We can't have the truck wait because we're offline," said Datre.

The company used ArcGIS Survey123 to scan a unique identifier to note the initial load, another app to measure the size of the load, and finally a third app to note the disposal and how the debris was separated by type to verify that the amounts matched up. The data from all three steps was then combined into a report.

"The initial challenge was that not only does it have to be disconnected but there's dynamically generated data throughout the day," said Phil Satlof, program manager with Blue Raster. It's not as easy as collecting the data offline and reconnecting online at the end of the day.

Blue Raster and CDR Maguire needed to determine where the

An excavator loads a truck in the early morning hours with wildfire debris that was chipped for mulching. Photo courtesy of CDR Maguire.

data would be kept, that it would be consistent across each form, and that it would be available when needed.

To start, each hauling truck had a unique ID that—once scanned into Survey123 with a QR code—would automatically populate the form with the name of the driver, area it was allowed to work, size, tonnage, and other attributes. The common ID then followed the load to inform anyone who touched it, conveying all these details by simply scanning the QR code.

The two companies had to figure out the workflow to account for changes such as new trucks being added and being able to append new records for each truck. To do so, a script was run in the background every 15 minutes to sync and stitch together matching records, pushing it to ArcGIS Online anytime internet access became available.

The new tool ensured that cleanup operations were not delayed, and Datre said CDR Maguire is looking at ways to further customize the solution to enhance efficiency.

The use of GIS for debris removal for emergency management has become vital, for effectively getting the job done and assuring the mobility of those affected.

"That visual perception of operations and progress is so critical," Datre said. "It's expected to see a map when you're asking, 'where is' the road cleared, for example."

A version of this story by Anthony Schultz titled "After Natural Disasters, CDR Maguire Innovates, Cleans Up Debris with GIS" originally appeared in the *Esri Blog* on October 4, 2022.

NEXT STEPS

A geographic approach to wildland fire

T HE MYRIAD PROBLEMS THAT WILDFIRES POSE REQUIRE A comprehensive solution. All stakeholders must collaboratively develop these solutions. The challenges also require continued preparedness and mitigation efforts and progress in response and recovery capabilities. Agencies can use an ArcGIS framework, the foundational location intelligence system that most national and state governments use, to perform data collection, analysis, operations support, and recovery efforts.

Here's how to get started using a geographic approach to wildland fire. For additional resources and links to live examples, visit the book web page.

Identify foundational data

You can gather and map foundational data in your area. These layers include basic infrastructure and administrative areas:

- Administrative boundaries (city/country boundaries, police districts and precincts, fire districts)

- Population and demographics

- Public safety infrastructure (fire stations, police stations)

- Vegetation type

- Slope

- Historic fire perimeters

- Structures and structure type (single-family homes, multifamily units, and commercial businesses)

- Major facilities and landmarks (schools, malls, places of worship, parks, and stadiums)

- Health infrastructure (hospitals, clinics, and assisted-living facilities)

- Shelters

- Roads

- Bridges

- Dams

- Utilities

- Communications infrastructure

- Water features (lakes, streams, and rivers)

- Parcels

- Addresses

The recommendation is to include hazard-specific data relevant for your area of interest. If you are unsure where wildfire presents the greatest risk in your area, tools such as ArcGIS Living Atlas of the World can help you identify risk and assess priorities. ArcGIS Living Atlas contains several ready-to-use live feeds that offer dynamic, real-time information that can be used in addition to your local data:

- Weather feeds

- Disaster feeds

- Earth observation feeds

- Multispectral feeds

Also consider adding real-time services for additional situational awareness:

- National Shelter System

- World Traffic Service

- Visible Infrared Imaging Radiometer Suite (VIIRS)

Identify data gaps

Once you have the base data collected and organized, you can identify data gaps. The data drill is a multiorganization exercise to gain insight into how a community collectively thinks about, manages, shares, and uses data during an emergency.

Data drills are developed and conducted based on operational challenges involving data and are a valuable tool for disaster preparedness. Data drills can be designed around a scenario relevant to your community, such as a flood, fire, or earthquake, to ensure you are planning for all data needs.

Here are a few things to consider in your data drill. Once the drill is completed, you can develop a plan to collect or create data where required based on these suggested steps:

- Detail your organization-specific operational workflows and use cases based on the scenario.

- Identify the relevant decisions that are needed and what datasets, including metadata and data dictionaries, support these decisions.

- Look next at your interagency workflows based on the scenario and identify the key decisions and support data needs.

- For each data point, identify the responsible organization contacts, roles, and responsibilities for this dataset.

- Identify whether partners will need any data-sharing agreements, and then start collecting and sharing the data identified in this drill.

Create and share maps

Once you've located the data sources you need, you can create a variety of maps to help make your community safer:

- **Map capacity:** Map your facilities, infrastructure, locations of employees or citizens, public safety resources, medical resources, equipment, goods, and services to understand your capacity.

- **Map threats and hazards:** Create maps showing the locations of highest wildfire risk and where evacuation routes and other preparedness and mitigation measures could help reduce risk.

- **Map vulnerable populations:** By mapping social vulnerability, age, and other factors, you can monitor at-risk groups and regions you serve, and that a wildfire could disproportionately impact.

- **Analyze the risk:** Create maps that analyze where threats and hazards intersect with facilities and the most vulnerable populations to identify areas at greatest risk. Then, analyze what is causing this risk by digging deeper into the data about the population, capacity, and hazards.

- **Share maps and plans:** Engage with your community to validate the data you are using to ensure an inclusive and collaborative process. By sharing your maps and plans, you help residents become part of the solution and ensure they also understand risks and increase readiness. This kind of collaboration provides transparency and improves community preparedness and compliance.

Follow best practices

To ensure that your maps and apps are ready to handle the load from the public and media and that your GIS environment is ready for the next response, follow the advice in the *ArcGIS Blog* post titled "Essential Configurations for Highly Scalable (Viral) ArcGIS Online Web Applications."

Learn by doing

Hands-on learning can strengthen your understanding of GIS and how it can be used to help address wildland fire. Esri provides a collection of free story-driven lessons allowing you to experience GIS applied to real-life problems:

- **Integrate maps, apps, and scenes to tell a story.** Share information about earthquake risk using maps, apps, and scenes.

- **Protect your home from wildfires by calculating defensible space.** Calculate the home ignition zone for a building in the fire-prone San Bernardino National Forest.

- **Automate fire damage assessment with deep learning.** Perform automated damage assessment of homes after the devastating Woolsey Fire.

- **Model landslide susceptibility.** Locate areas at risk of landslide damage after a wildfire.

- **Manage hydrant inspections.** Create and manage hydrant inspection assignments and complete them in the field.

Develop Community Wildfire Protection Plans

The Community Wildfire Protection Plans, or CWPPs, template can be used to facilitate the creation of a CWPP. These plans are collaboratively developed in the United States by state, tribal, and local governments; local fire departments; federal land management agencies; and other stakeholders. These plans help identify and prioritize areas for hazardous-fuels reduction treatments, identify response capacity, and document locations of residences and critical infrastructure.

Ask for help

Esri's Disaster Response Program (DRP) has provided GIS support to Esri users and the global community during disasters and crises of all types and sizes for more than 25 years. The DRP can provide data, software, configurable applications, and technical support for emergency GIS operations. Here's how the DRP can help:

- **ArcGIS software:** Existing Esri customers can temporarily extend existing licenses to support their organization's increased GIS requirements during a disaster response. If you're new to GIS or not yet a customer, you can gain temporary access to GIS software through the program.

- **Workflow implementation:** Esri can implement or help you configure solutions to support situational awareness, impact analysis, damage assessment, operational briefings, or public information during your response.

- **Data:** You can put your response in context by using existing data from ArcGIS Online and ArcGIS Living Atlas, such as real-time weather, traffic, hazards, infrastructure, and demographics. The DRP can connect you with incident-specific data shared by the response community and the private sector.

- **Technical support:** During your response, you can gain access to premium support services to address any question or issue related to ArcGIS.

- **Maps for the media:** You can request embedded maps or interview topic experts for disaster-related feature and news stories.

If you need emergency GIS help for your current disaster response, request assistance from the DRP. Visit esri.com/disaster for more information.

Learn more

For additional resources and links to live examples, visit the book web page at:

go.esri.com/prr-resources

CONTRIBUTORS

Jim Baumann
Chris Chiappinelli
Keith Mann
Amen Ra Mashariki
Monica Pratt
Ben Smith
Citabria Stevens
Carla Wheeler

ABOUT ESRI PRESS

ESRI PRESS IS AN AMERICAN BOOK PUBLISHER AND PART OF Esri, the global leader in geographic information system (GIS) software, location intelligence, and mapping. Since 1969, Esri has supported customers with geographic science and geospatial analytics, what we call The Science of Where®. We take a geographic approach to problem-solving, brought to life by modern GIS technology, and are committed to using science and technology to build a sustainable world.

At Esri Press, our mission is to inform, inspire, and teach professionals, students, educators, and the public about GIS by developing print and digital publications. Our goal is to increase the adoption of ArcGIS and to support the vision and brand of Esri. We strive to be the leader in publishing great GIS books, and we are dedicated to improving the work and lives of our global community of users, authors, and colleagues.

Acquisitions

Stacy Krieg
Claudia Naber
Alycia Tornetta
Craig Carpenter
Jenefer Shute

Editorial

Carolyn Schatz
Mark Henry
David Oberman

Production

Monica McGregor
Victoria Roberts

Sales & Marketing

Eric Kettunen
Sasha Gallardo
Beth Bauler

Contributors

Christian Harder
Matt Artz
Keith Mann

Business

Catherine Ortiz
Jon Carter
Jason Childs

Related titles

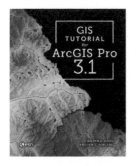

GIS Tutorial for ArcGIS Pro 3.1

Wilpen L. Gorr & Kristen S. Kurland

9781589487390

Top 20 Essential Skills for ArcGIS Pro

Bonnie Shrewsbury & Barry Waite

9781589487505

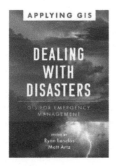

Dealing with Disasters: GIS for Emergency Management

Ryan Lanclos & Matt Artz (eds.)

9781589486393

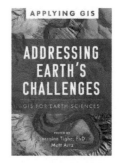

Addressing Earth's Challenges: GIS for Earth Sciences

Lorraine Tighe & Matt Artz (eds.)

9781589487529

For information on Esri Press books, e-books,
and resources, visit our website at

esripress.com.